SUSTAINABLE AGRICULTURE IN EGYPT

SUSTAINABLE AGRICULTURE IN EGYPT

edited by

Mohamed A. Faris
Mahmood Hasan Khan

Lynne Rienner Publishers ▪ Boulder & London

9-28-95

Published in the United States of America in 1993 by
Lynne Rienner Publishers, Inc.
1800 30th Street, Boulder, Colorado 80301

and in the United Kingdom by
Lynne Rienner Publishers, Inc.
3 Henrietta Street, Covent Garden, London WC2E 8LU

Library of Congress Cataloging-in-Publication Data
Sustainable agriculture in Egypt / edited by Mohamed A. Faris,
    Mahmood Hasan Khan.
    Includes bibliographical references and index.
    ISBN 1-55587-370-7 (alk. paper)
    1. Sustainable agriculture—Egypt. 2. Agriculture—Egypt.
3. Agriculture—Economic aspects—Egypt. I. Faris, Mohamed A.
II. Khan, Mahmood Hasan.
S473.E38S88   1993
389.1 ' 0962 —dc20   93-19486
                                                                CIP

British Cataloguing-in-Publication Data
A Cataloguing-in-Publication record for this book
is available from the British Library.

Printed and bound in the United States of America

The paper used in this publication meets the requirements
of the American National Standard for Permanence of
Paper for Printed Library Materials Z39.48-1984.

Contents

PART 3   DEVELOPMENT OF NATURAL AND HUMAN RESOURCES

PART 4   ADAPTATION OF TECHNOLOGY

PART 5   ECONOMIC ASPECTS

PART 6   CULTURAL, SOCIAL, AND POLITICAL ASPECTS

Acknowledgments

The publication of this book has left us in debt to several individuals and organizations who contributed in various ways to the international conference, "Sustainability of Egyptian Agriculture in the 1990s and Beyond," held in Alexandria, Egypt, in May 1992.

Our special thanks are first due to members of the Conference Organizing Committee for guidance and support long before the conference was held. The committee was chaired by H.E. Dr. Youssuf Wally, Egypt's Deputy Prime Minister and Minister of Agriculture and Land Reclamation. The committee's other members were Dr. Roger Buckland, Dr. Ahmed Momtaz, Dr. Mohamed A. Sabbah, and Mr. Aly Shady. The conference organizers were ably assisted by individuals from the Canadian International Development Agency (CIDA), McGill University, Alexandria University, and Egypt's Ministry of Agriculture and Land Reclamation. We are grateful to all of them.

The conference was attended by a large number of Egyptian scientists, academics, and policymakers, as well as international guests. We want to thank all of the active contributors to the seven conference sessions: keynote speakers, the presenters of twenty-five papers and those who took part in the discussions, and session rapporteurs.

We would like to express our thanks to Dr. Hassan Khedr, who spoke on behalf of Dr. Wally in the opening session. Our special thanks are due to Dr. Bekir Oteifa, who represented Dr. Wally in the closing session and made significant policy statements on the state of Egyptian agriculture and its prospects for sustainable growth in coming decades.

We also thank the contributors to the volume for cooperating with us in revising the original drafts and allowing us the freedom to make large and small editorial changes. Thanks, too, to Dr. Ahmed Momtaz for his helpful comments on the drafts, and to Ms. Anita Mahoney, who typed the drafts with patience and care. We also appreciate the cooperation of people at Lynne Rienner Publishers in getting the book published.

Mohamed A. Faris
Mahmood Hasan Khan

Introduction

Mohamed A. Faris and Mahmood Hasan Khan

This book contains edited versions of twenty-one papers presented and discussed at an international conference, "Sustainability of Egyptian Agriculture in the 1990s and Beyond," held in Alexandria, Egypt in May 1992. The papers address, mainly in the Egyptian context, the environmental, economic, cultural, social, and political aspects of sustainable development. In all cases, the notion of sustainability implies a central concern with the welfare of people, now and in the future, through increased productivity, for which the conservation and improvement in the quality of natural resources and the environment are the necessary conditions. The identified constraints on achieving and maintaining a sustainable agriculture, globally and nationally, range from individual attitudes and values to national and international policies.

"Sustainable development" has become more than a catch phrase since 1987, when the World Commission on Environment and Development (the Brundtland Commission) published its report, *Our Common Future*. The growing concern with both meeting the interdependent and changing needs of people and at the same time maintaining harmony with their natural resources and environment has started to yield concrete actions at national and international levels. It has also become evident that the issues of development and environment are contentious. The World Summit on Environment and Development in Brazil in June 1992 and the *World Development Report 1992* of the World Bank provided further evidence of the increased sensitivity to, and the difficult policy choices for, sustainable development at global, regional, and national levels.

The concept of sustainable development is inherently holistic: it implies a long-term perspective for planning, and integrated policies for implementation. As such its relationship to human welfare should take into account five basic dimensions: (1) environmental: maintaining the integrity of the bioecosystem in relation to human populations; (2) economic: increased efficiency in production and equity in the distribution of

1

resources and incomes; (3) cultural: sensitivity to the values and traditions of each society; (4) social: participatory development, including gender equality, with equitable access to social services and employment opportunities; and (5) political: empowerment, good governance, and protection of basic rights.

In the context of developing countries, the major challenge is to achieve and maintain a sustainable agriculture when agriculture is still the dominant sector of the economy. There is a legitimate concern in these countries about the future sustainability of agriculture. The needs of people for food and fiber have to be balanced with the needs of the natural environment and scarce resources of water and land. The challenge, in the words of the Brundtland Commission, is to develop agricultural systems "that focus as much attention on people as they do on technology, as much on resources as on production, as much on the long term as on the short term." This challenge takes on a special meaning in the developing countries, many of which are undergoing a painful experiment in "structural adjustment" to achieve sustainable development. Egypt is one of these countries, facing the dual challenge of making the painful structural adjustment in the short to medium term, and developing a sustainable agriculture in the long run.

Agriculture in Egypt

Historically, agriculture has been the dominant sector of the Egyptian economy. At present it produces nearly 20 percent of the gross domestic product (GDP); employs 37 percent of the labor force; contributes nearly 20 percent of export earnings. The country currently imports over two-thirds of its wheat and vegetable oils and one-third of its corn. Agricultural imports have increased threefold since the mid-1970s and now cost nearly US$3.5 billion per year.

Sustained economic growth in Egypt is closely linked to the prospects of growth in the agricultural sector on a sustainable basis, which contributes (1) to the welfare of the people in rural areas, about 40 percent of the country's population; (2) to consumption requirements (including food security) of a growing population with rising incomes and changing demand patterns; (3) to industrial production based on food and fiber; and (4) to earning/saving foreign exchange resources.

The agricultural sector, as did the rest of the Egyptian economy, operated from the early 1950s to the mid-1980s in a policy framework that was heavily influenced by the government. It was subject to distortions that adversely affected the performance of the sector, and it has received a declining share of public sector investment during the last twenty-five years. To alleviate these constraints, and ensure the development of sustainable agriculture in Egypt, public policy has responded in a variety of important ways since the mid-1980s. New investment projects have been completed in land and water resource development; others are under way in land reclamation and rehabilitation, modernization of irrigation structures, improvement in land drainage, control of waterlogging and salinity, increased efficiency of water-use on the farm, and the exploration of new water resources. Public investment is increasingly directed to improve the technology of farm production. Agricultural education, research, and extension services are being stepped up. At a wider level, policy changes have been made recently to improve Egypt's institutional and macroeconomic environment. These involve a gradual liberalization of domestic and foreign trade, deregulation of production, removal of price controls, and privatization of the production and distribution systems.

Major Issues

A sustainable development of the agricultural sector in Egypt is constrained by several factors. They include:

Scarce land and water resources: Egypt has a limited area of arable land suitable for crops. Added to this constraint are the pressure of rapid urban encroachments, degradation of land quality by intensive agriculture, and the onslaught of the desert. Even more important is the water constraint, because there is little potential for additional water and the demand is fast rising in both the agricultural and other sectors, particularly in industry and urban households.

Environmental degradation: Intensive agriculture based on limited land and water resources has led to considerable deterioration of land and water quality, which is exacerbated by the effects of high doses of fertilizers and pesticides, use of fossil fuels, and the increased industrial and urban waste contaminating the natural resource base.

Rapid population growth: Egypt's population is increasing at an annual rate of 2.7 percent and has so far shown no definite signs of a slowdown. Its effects on the economy and society are pervasive and generally deleterious.

Institutional arrangements: The land tenure system, including farm fragmentation, and the public sector control of land prices and rents have adverse effects on building farmers' organizations, adoption of new technology, and efficient use of land and water resources.

Macroeconomic environment: Public policies until 1986 were too interventionist in the process of private decision-making about allocation of resources, production, and marketing of agricultural products. Public investment in agriculture has neither been adequate nor emphasized its effectiveness or economic returns.

Agricultural administration: The support infrastructure for agricultural education, research, and extension has lacked effectiveness because of poor coordination and absence of direct participation by the end-users (farmers). Excessive bureaucracy and central controls have blunted incentives for farmers to organize and participate in making decisions at the local level.

While the agricultural sector in Egypt faces many serious challenges, it also possesses many untapped potentials. Recent policy changes have clearly shown that farmers would respond positively to opportunities made available to them. The potential sources for a sustainable process of agricultural development include:

- Improved management of irrigation water by adopting appropriate price and nonprice measures; e.g., formation of water-user organizations, pricing of water or cost recovery, regulation of water quality, and investment for increasing the efficiency of water-use.
- Increased yield-levels in the old lands through diffusion of improved technologies and cultural practices supported by a well-focused and integrated research and extension system. The potential for increasing the yield-levels in the new (reclaimed) lands (about 2 million feddans) may be even greater and can be realized by adopting similar measures.
- Improved agricultural administration, shifting (and not necessarily reducing) the role of the government from controlling the agricultural sector to providing an appropriate institutional (legal and regulatory) framework, an enabling macroeconomic environment, and infrastructural support for farmers. These will help to conserve scarce resources and increase efficiency, reduce the cost of production, and improve crop rotations and product-mix for higher returns and environmental protection. Government should decentralize authority and permit farmers to organize and participate in the key decisionmaking processes.

We must add that sustainable development of agriculture in Egypt would depend not only on the domestic challenges and opportunities but also on the international economic environment. Egypt's dependence on external factors has increased in recent years for several reasons. In this age of global interdependence between countries, Egypt, like many other developing countries, carries a large burden of foreign debt; faces a high level of agricultural imports; encounters trade and nontrade barriers for exports; and is going through a painful period of structural adjustment, including reduced spending, upward price adjustments, and unemployment. The process of adjustment has the support of the International Monetary Fund (IMF) and the World Bank.

There are serious concerns whether this structural adjustment will have a positive impact on the long-term sustainability of agriculture. The prospects of sustainable agricultural growth are expected to be affected significantly by the changes in international agricultural trade resulting from the Uruguay Round of negotiations in the General Agreement on Tariffs and Trade (GATT). At the World Summit on Environment and Development in June 1992, leaders, particularly those from the industrialized countries, did not make significant progress in acting concretely on issues that most urgently affect the world environment and the development prospects of developing countries.

Twenty-One Chapters: Different Perspectives

The Alexandria International Conference was called to focus on complex issues related to the ongoing efforts in the public and private sectors in Egypt to achieve sustainability in agriculture in the 1990s. The objectives were (1) to formulate concepts and principles for an integrated development of sustainable agriculture; (2) to analyze all aspects of sustainability—environmental, economic, cultural, social, and political—in the context of Egypt; and (3) to focus on future challenges and recommend actions for promoting the development of sustainable agriculture.

The edited papers in this volume reflect a common concern with the challenges and prospects for future sustainability of agriculture in Egypt. The authors' perspectives of course differ, depending on the personal or ideological viewpoint, the disciplinary approach or method, the issue under analysis, and the national (particular) or international (general) context. While some of the papers do not specifically address the problems in the Egyptian context, their analysis focuses on the central issues (challenges) related to the sustainability of the development process in agriculture.

The first three papers, chapters 1 to 3, emphasize the relevance of sustainable development to the agricultural systems of developing countries with rapidly growing populations, in the context of their goals of food security. Increased agricultural productivity is the key to achieve these goals—without compromising the long-term sustainability of the natural environment and resources for future (intergenerational) growth and welfare.

The general analysis by E.T. York Jr., in Chapter 1, focuses on three major constraints: rapid population growth, growing scarcity of water and arable land, and degradation of the environment and natural resources. It highlights the need to strengthen research institutions for integrating the needs of sustainability and productivity in agriculture. It is critical of the antiscience bias reflected in the "alternative agriculture" movements in developed countries like the United States.

The other two chapters in Part 1 are more specific about the problems and prospects of sustainable agriculture in Egypt in the coming decades. They reflect the so-called Non-Egyptian and Egyptian perspectives of two distinguished agriculturists, C.F. Bentley and Sayed Galal Jr. (assisted by Mazhar Fawzy), on the major issues of agricultural sustainability in Egypt discussed at the conference. These personal reflections highlight not only the issues but also a deep understanding of the complex process of change requiring trade-offs at the individual and societal levels. Also, they demonstrate the authors' sensitivity to the needs of present versus future generations; to balance between the needs of growth and the natural environment; to the requirements of increased productivity and equity, and the roles of market forces and the government in securing sustainability in Egypt's agriculture.

The environmental aspects of a sustainable agriculture are presented next, at a general level in Chapter 4 and in the Egyptian context in Chapter 5. Stuart Hill (in Chapter 4) contends that environmental sustainability is achieved only when individuals and societies use the renewable resources below the carrying capacity of the environment and the nonrenewable resources are conserved, recycled, and used mainly for emergency, and the level of impact on the environment is below its capacity to recover and continue to evolve. He differentiates between *shallow* and *deep* sustainability. The former takes for granted the goals of power and profit and makes little demand on the need to change values and lifestyles. Hill favors the deep sustainability paradigm, which requires three specific tools to redesign the agroecosystems: (1) education, research, and extension; (2) pecuniary incentives to adopt new methods and compensation for short-term losses in the transition; and (3) legislation and its enforcement, with penalties for those who persist in acting irresponsibly. The ability to achieve and maintain sustainability will rest—in Hill's view—on alternative visions and values, increased awareness and empowerment in all societies, and their appropriate use of science and technology.

Asit Biswas focuses on Egypt and the problems of water (and land). He emphasizes several issues of environmental sustainability with respect to irrigation water—the most important constraint on Egypt's agriculture—which he finds paramount in formulating a realistic agricultural strategy. He points out that the share of water for agriculture will decline in the coming decades; that conservation of water will play an increasingly important role; that dependence on water from underground sources will increase; that treated wastewater could be an important source of irrigation; that the issue of water quality has become quite serious because of increasing pollution; and that the existing legal framework against pollution should be amended significantly: pollution must be controlled at source.

The roles of natural and human resources receive special attention in chapters 6, 7, and 8. Abu-Zeid's view is that water, not land, is the major constraint for horizontal expansion of agriculture in Egypt. The high cost of land reclamation, increasing pressure of population, and higher standards of living will add to the demand for water—and water resources have little room for expansion. Major emphasis will have to be on efficient use of water resources, and major gains in efficiency can be made in the old agricultural lands, which depend almost entirely on surface irrigation. Abu-Zeid notes several constraints in this regard: their removal will have to include water charges or cost recovery from water users. Despite the potential problems and costs associated with re-using the treated and drainage water, these water management practices should be considered seriously and adopted. Of course, this will require continuous monitoring because of the possible impact on the environment.

Sustainable development of agriculture in Egypt will also depend upon the capabilities of the research and extension systems to develop or adopt appropriate technology and involve farmers in using it. In Chapter 7, Howard Steppler describes some of the conceptual and practical issues associated with this. Using a schematic cycle of technology generation—a research process involving five stages—he identifies the interdependent roles of agricultural researchers, extension agents, and educators. Their roles may differ in different stages of the process, ranging from specific to holistic. Steppler highlights the difficult problems associated with the development of an optimum curriculum for researchers and extension agents.

The issue of developing, adopting, and transferring new technologies is the focus not only of Chapter 7 but also of chapters 8 and 9. This section emphasizes the need to integrate several components into a viable system of research, training, and extension to make a substantial impact on both productivity and sustainability. Research and development should be oriented to solve the real problems of farmers and not to satisfy the personal or institutional idiosyncrasies of scientists and research managers. Hence, a need for farmer-participation in the research process. A strong, viable extension service should be linked closely to both researchers and farmers. These chapters also call for a well-developed delivery system, for inputs, relying on private markets with infrastructural support from the public sector. An efficient marketing system is also necessary to ensure a profitable return for farmers.

Sayed Balal and Hamdy Eisa, in chapters 8 and 9, focus on new technologies for rice and horticultural crops. These chapters examine Egypt's experience with rice and China's and Chile's with horticultural crops. Commercial production of new technologies (which include plant varieties) must be balanced with national priorities to conserve resources and minimize damage to the environment. Rapid increase in yield-levels may be obtained by using a lot of costly inputs that in the long run damage the environment.

Five chapters in the book deal with the economic aspects of agricultural sustainability in Egypt. A large part of the analysis in chapters 10, 11, and 12 focuses on policy changes that the Government of Egypt initiated in 1986, to be followed in 1991 by the structural adjustment program supported by the IMF and the World Bank. Ahmed Goueli (Chapter 10) identifies five phases of national agricultural policy from 1952 to 1992. He believes that, while liberalization of markets and removal of price distortions are necessary, they will not be sufficient to sustain agricultural growth in Egypt. Institutional reforms throughout the economy are also needed.

Goueli also emphasizes the need to reduce population growth, reduce dependence on food aid (and increase food security), improve water-use efficiency, and make the agricultural research and extension system more effective.

Ngozi Okonjo-Iweala and Youssef Fuleihan, in Chapter 11, tentatively analyze the impact of the adjustment program on the agricultural sector. They provide a background to the program, including the macroeconomic environment, its goals and basic ingredients. The initial signs seem to be positive: production has increased and the sector's net transfer to the rest of the economy has declined sharply. However, wage rates appear to have fallen in the last five years. Full liberalization of cotton and sugarcane is under way and so is the elimination of subsidies on inputs. The price of water has to be rationalized to discourage production of water-intensive crops.

Salah El-Serafy, in Chapter 12, takes a pessimistic view of the past performance of agriculture in Egypt. He cites both the annual growth rate and high input requirements for high yield-levels. The lackluster performance is due to factors such as low productivity of reclaimed lands; implicit taxation of the sector; pervasive government intervention; and rapid population growth. He warns that the structural adjustment program "per se may not take us along a sustainable growth path." He sees a possible inconsistency between the short-term aims of the program and the long-term needs of sustainable agriculture. The implications are that the government should help farmers optimize the use of scarce resources (especially water) within a macroeconomic environment of liberalized domestic and foreign trade. El-Serafy also emphasizes the need for deep institutional reforms—painful as they may be—during the transition to sustainability.

Joseph Potvin's analysis in Chapter 13 of factors external to Egypt surveys the country's dependence with respect to the flow of Nile waters. He also discusses vulnerability to global warming; and the external finance and trade requirement of "good environmental behavior." It may become increasingly difficult for Egypt, a trading and borrowing country, to observe the "acceptable" environmental standards. Potvin, like other contributors, draws attention to Egypt's population problem. He examines the high throughput of energy and raw materials relative to production levels and makes a controversial suggestion: to preserve the agroecosystem, Egypt should encourage widespread use of the organic methods of farming.

Maamoun Abdel Fattah has a narrow focus in Chapter 14: the impact of the Uruguay Round of GATT negotiations. Egypt—he points out—initially may have to face an increased import bill for food because of the elimination of agricultural subsidies in Europe and North America. With other developing countries in a similar predicament, Egypt is asking for food aid, technical assistance to enhance production, and improved market access in developed countries as fair compensation for the increased cost of imported food. It should be noted that Egypt's agricultural exports will at the same time enjoy new opportunities.

Sustainable development, however defined, rests on a complex and evolving interaction between humans and their environment/resources, and seven papers in the volume deal with cultural, social, and political factors. These factors are often contentious: they impinge on individual rights, values, and ethics; and empowerment by race, gender, and class.

The cultural aspects of sustainability are analyzed in chapters 15, 16, and 17. Michael Cernea, in Chapter 15, shifts the focus to a "culture of sustainability," emphasizing the role of behavioral norms and organizations in the private and public sectors. He contends that planned government interventions to induce development must work with culture, which is humanity's basic mechanism for adaptation to, as well as transformation of, nature. Social organization is the core of culture. Cernea maintains that it is necessary and possible to encourage social behavior conducive to environmentally sound production practices if they are built upon the heritage of good traditions, strengthening grassroot organization at local community level.

Nicholas Hopkins argues that any concern for agricultural sustainability in Egypt must take into account the centrality of the role of small farmers. To stress the role of small farmers is to underline the importance of the family and the household as the key organizational units in agriculture and rural life. A failure to understand the rural social organization, based on the small farm household, can lead to policies that destroy the fabric of rural life. This in turn can lead to an accelerated migration of people to urban areas and provide the social basis of a new regime—as has happened recently in other countries. Drawing on information from case studies and surveys, Hopkins reviews the technical and organizational aspects of the role of the small farm household.

In Chapter 17, Hoda Badran surveys the role and status of women in rural Egypt and suggests policy options to promote women's rights in the context of sustainable agriculture. She treats the status of women as a human rights issue. Women have made major contributions to agricultural production (a fact which is not well revealed by national labor statistics) and yet they suffer greatly at both household and community levels. A major change in national convictions and policies is required to make women an integral part of the issue of future sustainability of agriculture, Badran maintains. More of the same is not what is needed.

The social aspects of sustainability are addressed in chapters 18 and 19. Greg Spendjian believes that sustainable agriculture is unlikely to develop when a society's elites are not committed to the overall goals of sustainability. Simultaneously, the imperatives of economics, ecology, and ethics must be optimized. Human beings must become more purposeful and consciously goal-seeking; adopt long-term perspectives; and recognize global ecological and economic interdependence and connectedness. Spendjian believes the main sociopolitical question is: how will the costs for promoting conservation and enhancement of the natural resource base be met, and who will bear them? The answer implies tough political and social choices, tradeoffs, and even redistribution of wealth and resources within and between societies.

Ismail Sirageldin, in Chapter 19, discusses Egypt's population dynamic and the question of sustainability of current policies and trends. He maintains that Egypt is in the early stages of agricultural transformation, and he finds that the country's population response, especially in terms of geographic mobility, has been independent of the process of agricultural change. Population flow has not followed the pattern to be expected in an orderly agricultural transformation process. Sirageldin says this unsustainable process has been buttressed by public policy over a long period.

The final chapters, 20 and 21, analyze the political aspects of sustainability. Fahmy Bishay explains issues of national planning and policy analysis for an environmentally sustainable development of agriculture. He emphasizes the need to integrate the environmental dimensions into agricultural planning, using the planning-in-stages approach. He examines in detail major economic policy instruments that can create powerful incentives for the adoption of resource-saving and waste-reducing products and processes. In addition, he says, institutional reforms may be needed to reinforce the positive impact of public policies and markets.

Alan Richards, in Chapter 21, argues with some supporting evidence that Egypt has been living beyond its means for a long time. He says economic reform offers the only hope for job creation, food security, and poverty alleviation. The restructuring process, so necessary for sustainability, must shift most of the production to the private sector. But that does not imply a reduced role for the government: it should concentrate on those functions which it alone can fulfill. An important way in which the public sector can do its job better is for it to decentralize more and encourage citizens' participation. In Egypt, says Richards, the balance between private and public sectors has to be radically altered to facilitate long-term sustainability of the national economy and society.

Part 1

Major Issues for a
Sustainable Agriculture

1

Achieving and Maintaining a Sustainable Agriculture

E.T. York, Jr.

Concepts of sustainability have been applied in some disciplines for many years. However, the term *sustainability* has come into global use only over the past decade. In both developing and industrialized countries, during the 1980s there emerged a growing concern over the manner in which many of the earth's natural resources were being used and whether, with such usage, the needs of a steadily increasing population could be sustained. To put this concern in perspective, however, it should be noted that the twentieth century has seen remarkable progress in all areas of human endeavor. Education, medicine, industry, commerce, agriculture—all have advanced. These advances have resulted in better living conditions, increased life expectancy, better educational opportunities and higher literacy rates, improved food supplies, better nutrition, and a general improvement in the quality of life for many people around the world.

There is growing concern, however, that this progress may not be sustainable, because in making these advances we have exhausted inordinate amounts of nonrenewable resources; we have used, misused, and abused many of our renewable natural resources; and we have contributed to the degradation of many aspects of our environment in ways that could jeopardize the very future of humankind itself. While reflecting on this progress, it should be noted that millions of people around the world have not enjoyed this improvement in quality of living. The global community is, therefore, faced with a challenge of trying to improve the lot of those who have been largely bypassed by human progress and, at the same time, sustaining the progress made by others. Moreover, there is need to do this in ways that do not limit the abilities of future generations to enjoy similar or greater progress.

The Brundtland Commission

This challenge was the motivation for the United Nations to establish, in 1983, the World Commission on Environment and Development. The

15

commission, chaired by Prime Minister Gro Harlem Brundtland of Norway, was charged with formulating long-term strategies to achieve sustainable global development by the year 2000. In its 1987 report, *Our Common Future,* the commission defines sustainable development as "development that meets the needs of the present without jeopardizing the ability of future generations to meet their needs."

Applying this notion of sustainability to agriculture, one of the commission's panels stated that, "Enduring food security will depend on a sustainable and productive resource base. The challenge facing governments and producers is to increase agricultural productivity and thus insure food security, while enhancing the productive capacity of this natural resource base in a sustainable manner." The panel expressed this challenge in these words:

> The next few decades present a greater challenge to the world food systems than they may ever face again. The effort to increase production in pace with unprecedented increase in demand, while retaining the essential ecological integrity of food systems, is colossal, both in its magnitude and complexity. Given the obstacles to be overcome—most of them man-made—it can fail more easily than it can succeed (*Food 2000* 1987).

Especially note the emphasis that the Brundtland Commission places upon the need *to increase agricultural productivity.* This need for increasing output is often grossly neglected by many of those currently addressing the sustainability issue. They tend to place primary emphasis upon the issues of environmental or natural resource degradation—which are indeed very important issues—but essentially ignore the need for greater agricultural output.

A distinction should be made between sustainability and productivity. Greater productivity will be required to achieve the goals of sustainability, but it must not jeopardize agriculture's ability to meet future needs. Productivity goals achieved through short-term approaches may often not be sustainable, and efforts to achieve sustainability goals must take into account long-term implications and needs.

Horizontal and Vertical Expansion

In the second half of the twentieth century, the pattern of growth in agricultural production has changed. Before World War II, most of the increase in global production occurred as a result of expanding the area cultivated (horizontal expansion). As more production was needed, more land was brought into cultivation. The period after World War II has seen an unprecedented growth in agricultural production. Globally, agricultural output has grown at a rate of 2.5 percent per year. Moreover, this growth rate has generally exceeded the growth in population, resulting in an overall increase in per capita food production of approximately 0.6 percent annually between 1950 and 1986. This growth can be attributed not so much to horizontal expansion but rather to greater productivity (vertical expansion), resulting from the development and application of improved technology and related factors.

Problems of Hunger and Malnutrition

These statistics about the growth of world food output are deceptive: not all regions of the world have shared the experience of adequate food supply. In Africa, especially sub-Saharan Africa, for the past twenty years or so, per capita food production has declined at a rate of approximately 1 percent annually. Countries like Egypt have been the exception in that they have enjoyed a relatively high level of productivity: average crop yields compare favorably with production levels in the United States and other industrialized countries (York et al. 1982). Extensive areas of Asia and Latin America also have not met basic food requirements. Even in regions that normally do have good food supplies, temporary shortages and even famine can result from war, floods, droughts, earthquakes, and other disruptive disasters. The World Bank estimates that more than seven hundred million people, about one-third of the population in developing countries, do not receive enough calories for an active working life. This, in turn, is largely because they do not have adequate purchasing power: many of the world's hungry and malnourished are not able to buy the food that is available.

Future Food Prospects

If the recent impressive overall growth in agricultural production has fallen short of meeting global food needs, what are the prospects for the future? Current trends in food production do not offer great promise. Indeed, growth in agricultural production in much of the Third World is slowing significantly. In the thirty-six-year period from 1950 to 1986, in four of the six developing-country regions (North Africa, sub-Saharan Africa, and South and West Asia), the average growth in per capita food production was less during the last nine years of the period (1977–1986) than for the entire thirty-six years. These data suggest that many developing countries have fallen behind in efforts to meet the growing needs for agricultural products (York 1991). Egypt—it may be noted in passing— notwithstanding its rapid agricultural progress, is currently importing over two-thirds of its wheat and vegetable oils and one-third of its corn. Agricultural imports have reportedly increased threefold since the mid-1970s.

Since 1986, there have been sharp reversals in the gains made earlier in global cereal production. There were major declines in production in 1987/88 after slight increases in 1985/86. Brown (1988) states that in the mid-1980s grain production had stabilized in some of the world's most populous countries—India, Indonesia, Mexico, and China—that earlier had enjoyed impressive growth. Herdt (1988) and others have pointed to the closing gap between the average national yields of major food commodities and potential yields. In tests at the International Rice Research Institute (IRRI) in the Philippines, for example, maximum yields of rice have apparently not increased since 1965. Many believe that the Green Revolution—which saw remarkable progress in cereal production in the last two to three decades—has essentially run its course and future advances in agricultural output will depend on research breakthroughs in the development of production technology.

These trends are not encouraging, and dark clouds on the horizon suggest that the problem could become much worse. Evidence on three fronts supports this contention:

1. Population Growth: The demand for food is steadily growing as over ninety million people are added to the world population annually. More than 90 percent of this growth is occurring in the developing world, where serious problems of hunger and malnutrition already exist.
2. Arable Land: Another cause for concern is the growing difficulty in expanding areas of productive arable land well suited for cultivation. It is estimated that from 1975 to the year 2000, the area of cultivated land globally will expand by only 4 percent while the world population will increase by about 40 percent. Egypt, already using essentially all of its arable land, can expand agricultural production horizontally into the desert only at considerable cost.
3. Environmental and Natural Resource Degradation: A third and most disconcerting concern is the belief, held by many, that we are compromising the ability of future generations to meet their food needs by our current misuse of the natural resources on which agriculture depends. These problems include, among myriads of other difficulties: the rapid destruction of tropical forests; increasing concentration of atmospheric CO_2 levels and what some believe is the related global warming trend; destruction of the ozone layer in the stratosphere as well as ozone pollution problems near the earth's surface; major problems of soil erosion; contamination of water supplies; acid rain; and salinization and waterlogging of irrigated areas. Agriculture is viewed as a contributor to, as well as

a victim of, many of these global environmental difficulties.

The Issue of Agricultural Sustainability

In response to these problems, a concept of *alternative agricultural systems* has developed in the United States. It refers to agricultural systems that are alternative to the so-called conventional systems (National Research Council 1989). The U.S. Department of Agriculture has defined alternative agriculture as "a production system which avoids or largely excludes the use of synthetically compounded fertilizers, pesticides, growth regulators and livestock feed additives to the maximum extent feasible" (Ikerd 1989). Increasingly, the term *alternative agriculture* is being used to include what is commonly referred to as organic farming, regenerative agriculture, ecological agriculture, and low-input agriculture. Sometimes these alternative systems are equated with sustainable agriculture.

It seems that the basic concept of sustainability, as defined by the Brundtland Commission, is being significantly distorted by the manner in which the concepts of alternative agriculture are being equated with sustainable agriculture. The primary focus of the alternative systems is to reduce or eliminate the use of chemical inputs by replacing them with animal manures, crop rotations, and related practices. Many of the alternative practices have well-recognized merit, and there may be special circumstances in which they may be successfully employed. However, I question the feasibility or practicality of generally substituting most of these practices for those currently in use in highly productive agricultural operations in ways that can fulfill the twin objectives of productivity and profitability.

Many individuals and organizations in the United States are approaching sustainable issues on a much more balanced and substantive basis. They take into account not only environmental issues but also the economic viability of such systems. The American Society of Agronomy, for example, defines a sustainable agriculture as "one that over the long term (1) enhances environmental quality and the resource base on which agriculture depends, (2) provides for basic human food and fiber needs, (3) is economically viable and (4) enhances the quality of life for farmers and society as a whole" (Weil 1990).

The Research Advisory Committee (RAC) of the US Agency for International Development (USAID) addressed at some length the issue of low input and sustainable agriculture. In response to the contention that the modern or conventional agricultural systems were not sustainable, RAC stated, "Many modern agricultural production systems are not only sustainable, they have, in fact, created the fertility and resource base that sustains them. Some of the nation's most productive soils were once considered infertile and unproductive.... Most low-input systems require high labor input and are often characterized by low output" (Research Advisory Committee 1986).

Ikerd (1989) provides further perspective on this subject, suggesting that:

> A sustainable agriculture must be made up of farming systems that are capable of maintaining their productivity and usefulness to society indefinitely.... In the long run, farming systems must be productive, competitive and profitable or they cannot be sustained economically. Also systems must be ecologically sustainable or they cannot be profitable in the long run.

The Technical Advisory Committee (TAC) of the Consultative Group on International Agricultural Research (CGIAR) suggests that sustainable agriculture must involve the successful management of resources to satisfy changing human needs while maintaining or enhancing the quality of the environment. The goal of sustainable agriculture should be to maintain production at levels necessary to meet the increasing needs and aspirations of an expanding world population without degrading the environment (TAC 1989).

I have emphasized the fact that to be sustainable, agriculture must be made more productive to meet growing needs. Furthermore, to remain competitive and economically viable, agriculture must continue to be highly efficient and farmers must receive a reasonable return on their investment and labor.

Constraints on Sustainable Agriculture

There are many constraints to achieving the objectives of sustainable agriculture. Some of the more important are:

Soils and Water

Good quality soils, evidently of critical importance for crops, provide many of the nutrient elements essential for plant growth. Deficiency in these elements may seriously limit growth and productivity. An excess of nutrients and other chemicals can be toxic and can also limit crop production. Soils which are either too acid or too alkaline may impose further limitations on production. The physical characteristics of soils may also influence crop productivity by limiting aeration, root growth, and development.

Inadequate water is perhaps the single most important factor threatening the food security of many areas of the world. Nonsustainable use of water is occurring in many agricultural regions. There is much evidence that irrigation water is often used very inefficiently. In many areas, irrigation management agencies are unable to distribute water equitably, reliably, and efficiently. This has resulted in a reduction of the area effectively irrigated, lost productivity in areas poorly supplied, high cost of maintenance and rehabilitation of irrigation systems, and a general deterioration of the systems. Poor irrigation practices can result in severe problems of land degradation through waterlogging and salinization. The problem is particularly acute in India and Pakistan, where it is estimated that 12 million hectares have been seriously degraded. It is also a serious problem in Egypt.

There is a growing concern, particularly in industrial countries, about problems of water quality and pollution—especially from urban waste and from industrial and agricultural practices. Increased future demands for water from urban and industrial sectors could lead to a decline in the water resources available for agriculture. The problem will be exacerbated by the currently reduced rate of investment in new irrigation systems.

Atmospheric Pollution

Many human activities release harmful gaseous elements into the air. In some areas soils may already be too acid for optimal crop performance. This acidification may increase the solubility of certain elements in the soil to the extent that they become toxic to plants. The combustion of fossil fuels as well as the gaseous products of industrial processes may contribute to toxic ozone levels near the ground. Moreover, at higher altitudes some of these gases may damage the earth's protective ozone layer, increasing harmful ultraviolet radiation. I believe that the jury is still out with regard to global warming. While many indications suggest that global warming might be occurring, there is no conclusive evidence on this. If it should occur, there will be serious implications for agricultural productivity and sustainability.

Chemicals in Agriculture

Industry, and its products, may cause many hazardous chemicals to be released into the environment, threatening both the growth of agricultural crops and their quality. There is growing concern over chemicals in agriculture and related issues of environmental pollution and safety. Serious problems have been caused by the use of chemicals in agriculture. Illnesses and even deaths have been caused by the use of pesticides, particularly by people not applying the material correctly, in accordance with recommendations. Many problems of water pollution have resulted not from the recommended use of chemicals but from their misuse. We need to reexamine our recommendations with regard to fertilizer and pesticide use and make certain that we are not recommending more than needed. We need to encourage industry to develop fertilizers that are less subject to leaching. We also need to encourage industry to develop pesticides that are more rapidly degradable.

Food Safety

If agriculture is to be sustainable, it must provide safe and healthy food. There is much publicity today about the hazards of chemical residues on food, but evidence is growing that these hazards are not nearly as great as has been claimed. Many believe that such hazards are grossly distorted (Brookes 1990). For foods to be safe from pesticide residues, pesticides must be applied in accordance with recommendations and there must be adequate monitoring by regulatory agencies. People working in agriculture must do everything to reduce and, so far as possible, eliminate potential hazards. More research is needed with chemical inputs to determine optimum safe levels of usage.

Energy

Intensive production systems have high energy requirements and one of the important determinants of sustainability could become the availability of suitable sources of energy. Fossil fuels to operate tractors and irrigation pumps and the manufacture of fertilizers and pesticides account for most energy purchased by farmers. Those who contend that too much fossil fuel energy is being used in agriculture should note that, globally, this sector is only a modest consumer of energy. Agriculture accounts for only 3.5 percent of commercial energy used in industrial countries and 4 percent in developing countries. The strategy to double food production in developing countries through increased use of fertilizer, pesticides, irrigation, and mechanization would add only 5 percent to the present world energy consumption. This represents a small part of the energy that could be saved in other sectors through efficiency measures.

Crop and Livestock Pests

Intensification of production brings with it a greater risk of the build-up of pests. If not controlled, these pests affect stability of production in the short run and sustainability in the long run. *Pests* is used here to refer to weeds, diseases, insects, mites, nematodes, birds, and rodents and other mammals. It has been estimated that globally pests contribute to the loss in the field of some 35 percent of the potential production of major food crops, the greatest losses occurring in developing countries (Whitwer 1986).

The control of diseases and parasites is also important in sustaining livestock production. Globally, diseases and parasites are said to be responsible for the death of 50 million cattle and water buffalo and 100 million sheep and goats each year (Whitwer 1986). And these figures do not indicate the full extent of the problem: diseases and parasites may seriously reduce the productivity of animals without causing death.

Political, Economic, and Social Constraints

Many of the constraints on agricultural sustainability have their origins in the physical and biological environment, but there are also major problems which relate to the political, economic, and social environment. Political instability, for example, in some developing countries has been a major deterrent to sustained development. In the period from 1965 to 1980, of the sixteen African countries which failed to achieve an annual agricultural growth rate of 1 percent or more, thirteen of them had one or more major political crises during that period—civil war, invasion, or major coup.

In many developing countries, sustained agricultural development is hampered by the low priority afforded agriculture by national and local governments. The low priority is reflected in a low level of investment in the development of the agricultural sector and in trade, taxation, and pricing policies which often tend to favor the urban consumer at the expense of the farmer.

The Food and Agriculture Organization (FAO) suggests that an urban-biased development strategy has failed agriculture in many developing countries. Urban bias has not only been a deterrent to agricultural production, it has also reduced the demand for domestic products in relation to imported commodities. The domestic terms of trade are often sharply skewed against the rural sector. Overvalued currencies, for example, have made imports of food artificially cheap in relation to domestic production. Production levels of food can be significantly influenced by price levels established by government for food commodities. A good example of that exists in Egypt.

Availability of Inputs and Credit

The elimination of some of the major constraints on achieving sustainability will require the use of purchased inputs such as seed, fertilizers, pesticides, implements, and equipment. The lack of availability of these inputs at reasonable prices, as well as the lack of access to adequate credit to purchase them, can have serious effects.

The Brundtland Commission emphasized the need for more equitable access to land and water resources in developing countries. Land tenure arrangements may significantly affect natural resources. Many developing countries do not have adequate laws to protect agricultural lands from indiscriminate exploitation. Some countries have adequate legislation but have not been able to enforce it.

Research, Extension, and Education

Weak and ineffective research, education, and extension programs in agriculture usually cause a serious bottleneck in the development of sustainable agriculture. A recent World Bank Symposium emphasized the vital role improved technology must play:

> Improved technology development, diffusion, and adoption is at the heart of sustained agricultural advancement. Traditional agricultural systems, developed by trial and error over generations could be sustained indefinitely as long as the demands on the resources base did not exceed the rejuvenation capacity of that base. Increased population demands have disrupted this balance and technology must be continuously pumped into agricultural systems to sustain them above their natural state level (Hayward 1987).

Strong, effective research, extension, and educational programs are essential to the continued development, dissemination, and adoption of improved technology. This includes, at the national and international levels, the development of plant cultivars with improved characteristics such as resistance or tolerance to a wide range of diseases, insects and pests, and other adverse environment factors (Plucknett and Smith 1986).

Population Growth

Earlier I pointed to the rapid rate of population growth and the resulting growth in demand for food. No other single factor represents a more significant constraint on the achievement of sustainability objectives. Rapid population growth must be significantly slowed if we are to avoid the sort of chaos and misery—due to the inability of the world to feed itself—that Malthus envisaged some two centuries ago.

Research Implications

I have placed great emphasis upon the need for research to overcome the obstacles in the way of achieveing sustainability goals. What are the implications for research institutions? I would suggest that all research programs in agriculture should be planned, carried out, and evaluated with a sustainability perspective. We should constantly keep in mind the goals of sustainability and try to make sustainability concerns an integral part of all research efforts.

Unfortunately, in the United States there has developed a significant antiscience bias. This bias characterizes much of the alternative agriculture movement. I hope this is not true in Egypt.

An antiscience bias is truly unfortunate because the challenge of achieving sustainable agricultural systems rests in large measure with research and educational institutions. These institutions must focus increased attention on developing and applying the technology needed to achieve increased productivity and meet the economic and ecological dimensions of sustainability. The planet cannot achieve a sustainable agriculture and meet the evergrowing needs of people without the use of modern technology, including the appropriate use of agricultural chemicals.

References

Brookes, W.T. "The Wasteful Pursuit of Zero Risk," *Forbes*, April 30, 1990: 161–172.

Brown, L.A. *The Changing World Food Prospect: The Nineties and Beyond.* Worldwatch Paper 85. Washington, D.C.: Worldwatch International, 1988.

Food 2000: Global Policies for Sustainable Agriculture. London: Zed Books, 1987.

Hayward, J.A. "Issues in Research and Extension." A World Bank Symposium. Washington, D.C., 1987.

Herdt, R. "Increasing Crop Yields in Developing Countries." American Agricultural Economics Association Meeting. New York: The Rockefeller Foundation, 1988.

Ikerd, J.E. "Sustainable Agriculture." Annual Outlook Conference. Washington, D.C.: U.S. Department of Agriculture, November 29, 1989.

National Research Council, National Academy of Science. *Alternative Agriculture.* Washington, D.C.: National Academy Press, 1989.

Plucknett, D., and N. Smith. "International Cooperation in Cereal Research," *Advances in Cereal Science and Technology.* Vol. VIII. St. Paul, Minn.: American Association of Cereal Chemists, 1986.

Research Advisory Committee. U.S. Agency for International Development, Washington, D.C., 1986.

Technical Advisory Committee, CGIAR. "Sustainable Agricultural Production Implications for International Agricultural Research." FAO Research and Technology Paper 4. Rome, 1989.

World Commission on Environment and Development. *Our Common Future.* New York: Oxford University Press, 1987.

Weil, R.R. "Defining and Using the Concept of Sustainable Agriculture," *Journal of Agronomy Education*, 1990, 19: 126–130.

Whitwer, S.H. "Research and Technology Needs for the Twenty-First Century," *Global Aspects of Food Production.* Los Banos, Philippines: International Rice Research Institute, 1986.

York, E.T., et al. "Strategies for Accelerating Agricultural Development." A

Report of the Presidential Mission on Agricultural Development in Egypt. Ministry of Agriculture of the Arab Republic of Egypt and the U.S. Agency for International Development, 1982.

York, E.T. "Global Perspectives on International Agricultural Research." International Symposium on Physiology and Determination of Crop Yield. Gainesville, University of Florida, June 10–14, 1991.

2

Sustainability of Agriculture in Egypt: A Non-Egyptian Perspective

C. Fred Bentley

The conference theme—"Sustainability of Egyptian Agriculture in a Changing World"—recognized that there are challenges which must be faced. I think that three categories of challenges must be met successfully if sustainable agriculture is to be achieved and maintained in Egypt:

- Technical challenges: production, environmental, and economic problems
- Policy, administrative, and political challenges
- Challenges of traditions: problems of cultural and/or social traditions

All life depends on the environment of air, water, soil, and other biological organisms. Sustainability of agriculture depends on the use of these basic resources for agricultural production without degrading them; without, in fact, degrading a single one of them. No matter how marvelous technical advances and their applications may be, currently or in the future, there is a limit to the carrying capacity of Mother Earth. If we try to exceed that limit then agriculture, and human cultures as we know them, will not be sustainable: there will be enduring damage to one or more of the four fundamental components of the environment on which we all depend (Spendjian 1992).

Currently, a relentless increase in the demand for food is driven by population increase and economic growth. These pressures now endanger the future of humankind; hence the frantic search for sustainable agriculture. But there are many complexities. For example, successful coping with the water problem is dependent on meeting the technical challenge of using water more efficiently; the challenge that traditionally water has been free; and political challenges of accepting the necessity of implementing some system of charges on water before it is too late. Even if these challenges are met successfully, there is an ultimate limit—the carrying capacity—for sustainable agriculture in Egypt. In my opinion, there is dangerously inadequate understanding of the fact that we—civilization as now known—are approaching the carrying capacity of the world. We are impairing the environment on which we all depend. Sustainable agriculture is essential globally to the future: to our descendants.

When farming began, it became possible for some people to specialize. Education and accumulation of knowledge followed, and the technological marvels of modern civilization have resulted. But now, population increase, agricultural production, and industrial activities endanger the future of all people. Present trends—extinction of species, global reduction of biological diversity, pollution of air and water, and human-induced degradation of agricultural lands (as documented by the UNESCO-ISRIC global map)—are accelerating: in combination, these trends assure *unsustainability* of the environment—our life support—if they are continued. And, I repeat, all are accelerating.

Some may think these statements exaggerate the situation, trends, and prospects. The poisonous lead-laden smog over Cairo is now so serious that it is surely shortening the life-spans of Cairo residents, cumulatively, by tens of thousands of years, at least. The papers at the conference well documented Egypt's diminishing water availability per capita as well as the increasing amounts of industrial, biological, and agricultural pollutants in much surface and ground water. Development of additional land for agriculture is constrained severely by land quality and by formidable (should I say unsustainable) costs of development plus the high costs of maintenance of productivity under some current agricultural practices.

Major Challenges

The consensus at the conference was that water is the most important limiting constraint to sustainable agriculture in Egypt. It is clear that something must be done to improve the management, efficiency, and maintenance of the water-use systems. The imperative need for improved efficiency in use of irrigation water is made abundantly clear in papers by Biswas (1992) and Abu-Zeid (1992). In my opinion, the needed rate and amount of increase in irrigation water-use efficiency will not and cannot be achieved without introduction in the near future of universal and substantial charges for water. I think appropriate water charges for all other users should also become universal simultaneously, along with high additional charges on those who pollute water.

In view of the increasing scarcity of irrigation water, the introduction of universal charges for water should be a high priority for Egypt. Implementation of water charges should be based on very careful planning and preparations, with provisions for effective administration. To be accepted and effective, much of the management, administration, and operations for water-use must be by local communities, not by a bureau-cracy in Cairo (Bishay 1992). Such changes will cause some upheavals. But such actions appear to be essential for Egyptian survival.

Presentations and discussions at the conference concluded that land is agriculture's second most important constraint. As a soil scientist, I must emphasize a fact related to the real costs of the "new lands" that I think has not been made clear. Globally, there is a scarcity of new lands reasonably suitable to bring into agricultural production. In almost all cases, either the costs of development, or the inferior quality of such lands, or combination of the two factors, make the cost per unit of agricultural production higher on new lands than the costs on lands already in use. That is why financial institutions are unwilling to make loans for land development projects. That is also why few developing countries can themselves afford the cost of developing new lands. In most cases, inter-national development assistance agencies would be better advised to focus their assistance on increasing production on lands already being farmed than on new land development projects.

I would like to see a rigorous assessment of the real costs of develop-ment of new lands for irrigation. Such an evaluation should be done by competent, independent, nonengineering personnel. Items which should be included are costs of planning and monitoring; dams, reservoirs, and canals; infrastructure and water-control devices; irrigation equipment and farm machinery for the new farms; maintenance, operating, and supervi-sion services; an appropriate allowance for the generally inferior yields (or higher production costs) on new lands; and determination of the total cost of all of the associated corruption in both donor and recipient countries. I doubt that, with this cost assessment, development of new lands would be economic or defensible compared with the alternative of increasing production on lands already in use.

Before new land development projects are undertaken in Egypt, there should be realistic economic evaluations to determine whether larger, less costly, and sustainable agricultural production increases can be attained by investments in improved utilization of existing, under-exploited tech-nologies for water management and production of crops. I urge that such evaluations not be done by engineers or personnel engaged by them: the record shows that few engineers have satisfactory knowledge, apprecia-tion, or understanding of agricultural production, soils, biological things, and farm people.

Human Resource Development

The gross domestic product (GDP) of a country is determined by the productivity of its work force. In Egypt, nearly 40 percent of workers are employed in agriculture, which contributes only 20 percent to GDP.

Clearly the productivity of the agricultural labor force is far below the nation's average. The productivity of workers is strongly affected by their levels of education and training. So, productivity of agricultural workers, especially girls and women, is adversely affected by the inferior levels of their education. Girls and women are an important part of Egypt's agricultural labor force and their low levels of training limit their contributions to increased production.

Improved sustainable agricultural production is strongly dependent on use of improved production practices. But the ability of workers to adopt better methods of production is influenced by their levels of practical education. So, increase in the productivity of the agricultural labor force is a key to national welfare. Human resource development means increasing human capabilities by training and education. Globally, the employability of illiterate or untrained people is declining. Their ability to increase their productivity, and so their quality of life, is very limited; so too is their future. In developing countries, the methods and costs of training and education appropriate to an increasingly technical world pose difficult challenges. In this respect, Thailand has developed a highly practical solution. It has imposed a heavy tax on imported assembled motor vehicles. However, there is only a very modest tax on automobile components that are imported for assembly in Thailand. The assembly work must be done by Thais who are trained by the companies concerned. Thus, Thailand is rapidly developing a large number of highly skilled workers capable of applying their talents in many diverse industries.

Egypt would do well to explore opportunities to follow the Thai strategy. For example, sprinkler- and drip-irrigation is likely to expand rapidly in Egypt. Currently all such equipment is imported. Why not require a progressive increase of production in Egypt of components of such devices until such time as complete manufacture of them is done there? Similarly, why not require the companies that produce such devices to establish facilities in Egypt to train Egyptians to do such servicing and repairing? Some companies might object: let them; other companies will seize the opportunity to establish outlets in Egypt.

Another example of a strategy for human resource development: What is reported to be the world's largest (and very successful) cooperative began in Gujarat State, India, with the purchase of small quantities of milk (as little as one liter) from poor village women. These women received cash payment at once. Success and rapid growth of the cooperative led to the training of women about the nutrition and management of their animals. That started a chain reaction of various human resource improvement projects that have been very beneficial to poor women and their families.

Badran (1992) has documented that in Egypt, women—especially rural women—are not accorded equitable treatment and opportunities.

Cultural traditions are probably the most severe constraint to more equitable opportunities for women. However, it is very difficult to cope with the challenges of traditions. Fortunately, some donors are now willing to support programs designed to improve the status and conditions of women, especially rural women. It is hoped that women's organizations will develop proposals that will exploit the good will of donors who are anxious to be helpful.

Improved Technology Transfer

Improved and sustainable agriculture depends on effective use of knowledge—both existing knowledge and new knowledge as it emerges. The world is becoming increasingly technical, with many kinds of specialization. Therefore, interdisciplinary cooperation is necessary to exploit new knowledge for how best to produce a particular crop. A decade ago, a USAID/World Bank study reported (CIDA 1989) that production of cereals in Egypt could have been increased by an estimated 50 to 70 percent by more effective use of technologies that were available in the country at that time.

A similar situation occurred in a populous Asian country that sought massive development assistance to enable 50 percent increases in production of several crops within a decade. Part of the request was for much overseas graduate study for M.Sc. and Ph.D. degrees in research disciplines and extension. After an evaluation visit, an expatriate reported that he thought the country already had the capability to increase production of his crop-specialty by 50 percent if the knowledge he had encountered was used effectively. That statement sparked prompt action. Fifty on-farm comparisons between farmer practices and recommendations from researchers were conducted. Farmers did all of the production work on their own farms under the guidance of extension officers who had been trained for two weeks by researchers. Results of the average yield for fifty farms were: farm practices, 800 kg/ha; researcher's recommendations, 1,800 kg/ha—or an increase of 125 percent. Three fundamental points emerge:

- The researchers in various disciplines were working in complete isolation from each other; the directors-general of research, production, and extension had never worked together before on anything.
- The project requested was not needed. What was needed was interdisciplinary, interdepartmental, interagency cooperation—with all of them working together with farmers.
- The request for graduate training overseas in agricultural exten-

sion was found to be unnecessary.

Lacking intimate familiarity with the agricultural extension (technology transfer) situation in Egypt, instead of making comments, I will ask some basic questions:

- Can Egypt justify huge investments in land and water development if there already exist underutilized production technologies with the potential, if exploited, to increase agricultural production greatly and at very low cost?
- If it is contended there is no proof that large increases are possible by the application of existing, affordable knowledge, are there reliable scientific data (from appropriate on-farm experiments) to prove that existing recommendations are being employed by most farmers?
- Are the twenty-three thousand agricultural extension officers in Egypt competent and willing to assist and encourage farmers to conduct meaningful comparisons of their own practices against the recommendations of researchers?
- Is there adequate interdisciplinary cooperation within the agricultural universities? Between agricultural research groups or disciplines? Between research personnel and extension personnel? Between all of the above groups and farmers?
- Are agricultural students, professors, researchers, and extension agents appropriately familiar with farm problems? Are they comfortable and confident talking to farmers?

From experience in other countries, I know that too often there is little or no cooperation between the various possible combinations of research stations, universities, and extension departments working together with farmers on the lands of farmers.

If Egypt does not have a comprehensive program of on-farm experiments normally planned together by farmers, interdisciplinary researchers, and extension personnel, I recommend to the government and development assistance agencies that a carefully planned program should constitute part of Egypt's strategy to develop sustainable agricultural production.

Administration of Agricultural Programs

If user charges are established for irrigation water, Egypt will have income to pay the maintenance and operational costs for water distribution to farmers, and with improved efficiency. That should be an important improvement. But there is a problem: should the administration of those activities be from the top down or from the bottom up? Currently the role of village organizations is very small. It is doubtful that increased sustainable agricultural production will be attained without increased powers and roles at the village level.

Sri Lankan experience may be helpful in considering that question. For centuries, each local community in Sri Lanka employed a member of the community to maintain and manage the water distribution system for the village. Results were very satisfactory: the systems were kept in good repair, and the water supply was fairly and reliably available to users. But about twenty-five years ago a government with a commitment to central planning and administration took over responsibility for water distribution and maintenance of the systems. Within a decade, the systems were seriously deteriorated and there was deep dissatisfaction with the government.

Egypt should try to field-test irrigation organization, management, and maintenance, giving a major role (or full responsibility) to local people. Such testing might be tried at a few widely separated locations, with expansion to follow if it is successful.

Input Subsidies, Price Controls, and Quotas

Increased agricultural production does not always help rural people, relative to urban populations. Thailand, for example, has increased agricultural production dramatically during the past three decades. The Green Revolution there has sparked a remarkable economic revolution: Thailand ranks in the top ten countries in the world in sustained rate of economic improvement for the last ten years. The increased production per agricultural worker and per hectare has resulted in a decrease in the proportion of the Thai labor force employed on farms; the resulting increase in the urban work force has contributed to the rapid rate of economic improvement which has benefited all Thais.

But urban people, not rural people, have benefited most from economic improvement in Thailand. As described by the Thai minister of agriculture and cooperatives, in the 1960s the relative income of nonfarm people was slightly over three times that of the rural population. By the late 1980s, the relative income of nonfarm people was over six times that of farm people (Bentley 1991). Improved production per hectare and per agricultural worker increases per capita farm production and benefits the nation. But government price controls, quotas, and lack of competition in the marketplace keep the producer prices very close to the cost of production. Urban people receive more and better food at lower relative cost: they benefit greatly. Real incomes of farm people increase comparatively less, so the nonfarm people benefit most from improved productivity and efficiency of the agricultural producers.

Egyptian agricultural subsidies were supposed to benefit farm people, but there have been other effects, too. It was argued at the conference that price controls offset the advantages of input subsidies; that they have encouraged wasteful use of fertilizer and pesticides. Conference contributors stated, without citing proofs, that the subsidy-induced use of fertilizers and pesticides has degraded and polluted Egypt's agricultural land base and water resources.

In Canada, too, massive agricultural subsidies, primarily to grain producers, are causing problems. We produce more grain than we need for domestic use. But on international markets, Canada cannot sell all of its surplus grain at prices that cover the real costs of production, so food aid to developing countries has become a way of disposing of surplus. However, food aid from Canada keeps wheat prices to Egyptian wheat producers low. At the same time, the residents of Alexandria and Cairo benefit from low-cost wheat products. Meanwhile, subsidies in Canada are encouraging farm practices which are significantly degrading farming areas. Farmers and the sustainability of farming are the losers.

Fear of Agricultural Chemicals

I agree with York (1992) regarding the inappropriateness of the USDA description of the so-called alternative agriculture. To me the inclusion of the words "synthetically compounded fertilizers, growth regulators and livestock feed additives" in its definition is both antiscience and unintelligent pandering to the irrational public paranoia and fear which is so ardently fanned and promoted by some misguided evangelical environmentalists. Appropriate use of those materials is essential to maintain a highly productive and sustainable farming for a significant part of world agriculture. A global halt in the use of "synthetically compounded fertilizers to the maximum extent feasible" would decrease world food production so much as to cause the greatest crisis in human history.

There are fears about the appropriate use of rigorously evaluated and intelligently used pesticides. I will draw an analogy. Many people have benefited from medicines, vaccinations, inoculations, iodized salt, and a host of other synthetic chemicals—substances which they have ingested or which have been injected into them. Would the antiagricultural chemical lobby "decrease the use of *such* chemicals to the maximum feasible extent"? (My italics.) More than 99 percent of all pesticides in foods eaten by people are natural poisons—pesticides that plants produce and that protect them from pests. Sometimes foods contain seriously harmful amounts of natural pesticides or chemicals.

It is important, though, to adopt and enforce strict regulations, and great care should be employed to maximize the benefits from the appropriate use of approved agricultural chemicals. Use and sale of drugs and chemicals for health protection, or the saving of lives, are very strictly controlled: many of them are highly dangerous if misused. Why not regulate agricultural chemicals with comparable care and rigor?

Land Fragmentation and Community Lands

The continuing cutting up of small farms into ever smaller, less efficient, and scattered parcels causes significant inefficiencies of production and labor use. Egypt, more than many other countries with the same problem, cannot afford such waste of land and human effort. Land is too scarce and fragmentation can cause productivity to stagnate.

Appropriate action in this thorny area can bring important benefits. In India I have seen a 200-ha area where consolidation of the holdings of many owners, and a carefully planned layout, had these effects:

- Each owner's fragments of land were replaced by a single, conveniently shaped working area.
- All owners received uniform and improved irrigation.
- The potential was created for tractorization.
- The overall cultivated area was increased by two percent.
- Grain production increased by about 50 percent.
- Labor requirements decreased appreciably.

Community lands for use by all members pose a different kind of land management problem. Typically, common community-used grazing and treed areas are seriously depleted or desertified due to lack of management and gross overexploitation. Production from such areas is low and inefficient. Off-site areas often suffer damage because of erosion and runoff from such denuded lands. If there are such misused lands in Egypt, the using communities should be assisted and required to establish local management committees.

Seed System

At the International Institute for Tropical Agriculture (CIAT) in Cali, Colombia, a so-called seed unit was developed to increase the number of new crop varieties, from breeder's seed to quantities for distribution to farmers. The seed unit established rigorous procedures and high standards for the improved varieties, developed by CIAT and national agricultural researchers in Latin America. A former director-general of CIAT, John Nickel, said, "At the outset I had deep doubts regarding the need for the seed unit. Now, half a dozen years later, I think the work of the seed unit may be the most important contribution of CIAT to Latin America."

Nickel stated that as a result of training courses conducted by the CIAT seed unit, and the in-country promotion and assistance to seed units, almost all Latin American countries have, and use, high-quality seeds of improved, high-yielding varieties.

A survey of eight African countries in 1986 found generally unsatisfactory use of improved high-quality seeds by farmers (Bentley 1986). However, two countries had outstanding seed systems for hybrid corn and a few other crops. High-quality hybrid corn seed was available to farmers who needed as little as only 2 kg of seed. In both countries, the high-quality seeds were produced by self-disciplined seed-grower cooperatives. The farmers (who multiplied high-yielding variety seeds or produced hybrid corn seed) employed the management and workers of their seed units. Employees were given clear rules that had to be met or the workers would be released at once. They were to reject every field that failed to produce the approved high-quality seed. Any farmer whose seed production failed to meet all requirements of the frequent intensive field inspections, or whose seed on delivery to the seed plant failed to meet the quality standards, was unable to sell seed of guaranteed quality that year. When the sale of new seed for the next planting season began, hundreds of small farmers waited in long lines to purchase packets of corn seed.

The Population Issue

The rapid rate of population increase in Egypt, if continued, will probably be the most important impediment to achieving sustainable agricultural production. At the conference, the adverse effects of population increase were mentioned frequently, but few specifics were developed. Statements made at the conference included:

- Few countries are as close to the carrying capacity of their water and land resources as is Egypt.
- To create new jobs and hold wages and unemployment at present levels, Egypt will have to invest huge amounts. That will be an unending need as long as population continues to increase rapidly.
- Due to population increase, the gap between high-income and low-income groups may widen.
- Sustainable agriculture may be impossible if the current high rate of population increase continues.

But no one at the conference voiced a recommendation for action on the population problem. I do so, cautiously but strongly. Egypt should pursue a vigorous and strongly supported program of education, support, assistance, and incentives to encourage people to limit their family size.

Four Challenging Opinions

The following four personal opinions have, I think, very important implications for sustainable agriculture in Egypt:

1. Global environmental degradation is more serious and more endangering to the future than the public realizes.
2. In a finite world with resource limitations, the unending population increase and the polluting economic growth process are unsustainable.
3. Economic disparities within countries are growing. If such disparities are not reduced by actions of national governments, and rather soon, I predict unprecedented social anarchy or chaos in many countries.
4. We know that an uncontrolled cancer destroys its host. Rapid population increase is a cancer on Mother Earth which, if uncontrolled, will destroy the environment.

References

Abu-Zeid, Mahmoud. "Egypt's Water Resource Management and Policies." International conference, "Sustainability of Egyptian Agriculture." CEMARP, Alexandria, Egypt. May 1992.

Badran, Hoda. "Women's Rights as a Condition for Sustainability of Agriculture." International conference, "Sustainability of Egyptian Agriculture." CEMARP, Alexandria, Egypt. May 1992.

Bentley, C.F., R.G. Griffiths, and G.A. Reusche. "Improved Seed Systems for Africa." A Consulting Report to Winrock International. Arlington, Virginia. 1986.

Bentley, C.F. Summary of international workshop, "Evaluation for Sustainable Land Management in the Developing World." Chiang Rai, Thailand. IBSRAM, Bangkok, 1991.

Bishay, F.R. "Integration of Environmental and Sustainable Development Dimensions in Agricultural Planning and Policy Analysis." International conference, "Sustainability of Egyptian Agriculture." CEMARP, Alexandria, Egypt. May 1992.

Biswas, Asit K. "Environmental Sustainability of Egyptian Agriculture: Problems and Perspectives." International conference, "Sustainability of Egyptian Agriculture." CEMARP, Alexandria, Egypt. May 1992.

Canadian International Development Agency (CIDA). "Arab Republic of Egypt: ISAWIP." CIDA Project No. 344/11872. A Consulting Report by Dario Pulgar Communications. Ottawa, March 1989.

El-Serafy, Salah. "The Agricultural Sector in the Context of Egypt's Structural Adjustment Program." International conference, "Sustainability of Egyptian Agriculture." CEMARP, Alexandria, Egypt. May 1992.

Okonjo-Iweala and Youssef Fuleihan. "Structural Adjustment and Egyptian Agriculture: Some Preliminary Indications of the Impact of Economic Reforms."

International conference, "Sustainability of Egyptian Agriculture." CEMARP, Alexandria, Egypt. May 1992.

Richards, Alan. "Food, Jobs, and Water: Participation and Governance for a Sustainable Agriculture." International conference, "Sustainability of Egyptian Agriculture." CEMARP, Alexandria, Egypt. May 1992.

Spendjian, Greg. "Sustainable Development Necessitates a Social Revolution." International conference, "Sustainability of Egyptian Agriculture." CEMARP, Alexandria, Egypt. May 1992.

York, E.T., Jr. "Achieving and Maintaining a Sustainable Agriculture." International conference, "Sustainability of Egyptian Agriculture." CEMARP, Alexandria, Egypt. May 1992.

3

Sustainability of Agriculture in Egypt: An Egyptian Perspective

Sayed Galal, Jr. and Mazhar Fawzy

The concept of sustainability in the context of agriculture in Egypt, as defined at the conference, implies a dynamic process that, while providing economically the agricultural needs of the society, enhances the quality of human life, environment, and agricultural resources. A consensus about the desirability of sustainable agriculture does not, however, lead to an agreement on the tradeoffs involved in its feasibility.

The conference focused on two basic aspects of a sustainable agriculture in Egypt: *environmental*, including natural resources, population growth, agricultural inputs, and technology; and *political*, including the economic and social conditions and policies affecting agricultural growth and the quality of life of rural people. We will first survey these issues—with our "Egyptian" perspective on the problems—and then offer suggestions for a reorientation of public policy in several interrelated areas.

Environmental Aspects

There is basic agreement that increased agricultural production and productivity are the keys to maintaining economic sustainability in Egypt. Likewise there is strong evidence that most of the future agricultural growth will have to depend on better use of existing natural resources, particularly water and land. The Nile water system and underground water for irrigation are not expected to make significant additions to the existing supply, so the emphasis will have to be on improving the efficiency of water-use. Wastage of water and deterioration of its quality, and spoilage of land by waterlogging and salinity, are indeed the key constraints to the sustainability of agriculture in Egypt. Additional land for agricultural purposes will not be available at a reasonable cost, as past experience of land reclamation projects has clearly shown. The more important point is to focus on conserving the productive quality of the already cultivated area and to assure an adequate supply of the productivity-enhancing inputs, particularly irrigation water. Policies that encourage increased productivity of the existing water and land resources will also promote the cause of the environment.

It has become evident that horizontal expansion is not a feasible route for future agricultural development because of the economic limits on the availability of water and land. But vertical expansion by intensifying the use of agricultural inputs on the given natural resources has not been rationalized either. In fact, the singleminded pursuit of high-input intensive agriculture, particularly in the Nile Valley and the Delta, in the last two to three decades has produced serious environmental and economic impacts on the country. New technologies and products have not necessarily been cost efficient. The high yield-levels of major crops have been achieved without much concern about their costs in terms of resources (inputs) and the environment.

The natural resource base of Egypt is not only limited but its quality has deteriorated because of population growth and inefficient use of the key inputs: water, energy, and chemicals (fertilizers and pesticides). Some of the most fertile areas have lost their productivity, becoming increasingly unsuitable for crop and livestock production. Rapid population growth and urbanization have led to increasing encroachment on agricultural lands and the pollution of land and water resources. Slow and uncertain growth of agricultural incomes has been a major contributing factor in this regard. The short-term horizon of the small farmer (wishing to maximize returns) and public policy (guided by a similar horizon) have not been of much help either.

Agricultural sustainability has been threatened also by the way in which the key inputs for growth of productivity have been abused. Public policy in the past may have played an important role in this process. Expensive investments in the development of water and reclamation of land were not accompanied by policies emphasizing cost recovery, resulting in wastage of water and land and deterioration in the quality of the resource base. Similarly, big subsidies were given to promote the use of fossil fuels (oil) and chemical pesticides and fertilizers—without similar emphasis on developing a strong research and extension support system.

A perverse pricing system—which on one hand subsidized the cost of key inputs and on the other taxed farmers on their crops—combined with the weak support system led to a distorted allocation of resources, slow growth in incomes, resource degradation, and environmental pollution. Realistic penalties were not imposed on overusing the inputs that contributed to environmental and resource degradation. Profitable alternatives were not presented to farmers to rationalize the use of these inputs: e.g., new crop rotations, integrated pest management, or change in the mix of organic and chemical fertilizers. Now, the structural adjustment process, which started in the agricultural sector in 1986, is correctly addressing some of the pricing and control policies to promote sustainable agriculture in Egypt.

Political Aspects

Political aspects include all of the socioeconomic issues related to the sustainability of Egyptian agriculture. At the conference, two themes dominated: the role of public policy; and increased participation by farm households (small farmers in particular) in the process of decisionmaking at local and national levels.

The political and economic roles of the governments in Egypt, at least since the Revolution of 1952, have been pervasive but not always benefi- cent. Governments have introduced structural reforms affecting the own- ership and control of land; cooperatization of farmers; settlements projects on the new lands; allocation of land and water to different crops; accessibility to key inputs like water, fertilizers, pesticides, and credit; delivery of farm produce to markets; and import and export of commod- ities directly related to the agricultural sector. In addition, they have maintained a support system of agricultural education, research, and extension services to promote new technologies. All this activity, in both economic and social spheres, has affected the lives of millions of farm families.

These policies were controversial and hotly debated, but there was little focus on changes affecting the farm families with regard to their incomes and employment; size and composition of the rural household; the role and status of rural women; and rural to urban migration. There is considerable evidence that farmers (especially the *fellahin*) were by and large passive actors in planning, developing, and implementing most of the policies and projects affecting their incomes and jobs. Governance without participation has proved in many countries to be an inappropriate strategy for achieving a sustainable agriculture. Policies and schemes that do not emanate from those expected to benefit cannot address their problems and meet their basic needs. Bureaucracies without accountabil- ity to the people and their interests tend to breed inefficiency and rent- seeking behavior and this happened in Egypt until the government decided, in 1986, to start decentralizing the decision-making process, liberalizing its controls on crop areas, prices, and so forth.

There is now greater appreciation by the government of the fact that a sustainable agriculture needs two key elements from the government: (1) a *macroeconomic and legal framework* which produces the right signals for people to take risks, allocate resources rationally, and internalize (positive and negative) externalities; and (2) a *support system of physical and social infrastructure*—irrigation, roads, transport, communications, market information, and agricultural education, research, and extension services. Investments (in some of the physical infrastructure, and in almost all of the social) have a high social and economic return. The political feasibility of such investments depends on how well governments act in the interests of the farmers, allowing them to make most of the decisions, or at least participate in all stages of the decision-making process.

What Should Be Done in Egypt?

Given our personal perspective on the problems facing Egypt's agriculture, and in the light of the information made available at the conference, we want to make several recommendations on the major issues.

- First, clearly to define agricultural sustainability, we are comfortable with the definition given at the opening of this chapter. It implies environmental and resource protection (so that future generations are not imperiled) and satisfaction of the economic needs of the society (meeting the criterion of relative scarcity) with improvements in the quality of life. These two implications (environmental and economic) reinforce each other.
- We must develop monitorable indicators or measures of sustainability. This is linked directly to the way in which we make operational the definition of sustainable agriculture. Some variables can be measured directly; others may need proxy variables to reflect changes. It is also essential that the legal framework integrates these measures and that the regulations are enforceable at a reasonable cost. The "cost recovery" and "polluter pay" principles should be translated into concrete rules and regulations that are administratively efficient and cost effective. Many existing laws and regulations are neither efficient nor effective in their impact on people's behavior.
- It is essential to involve farmers and other relevant parties directly as decisionmakers and intended beneficiaries. Decentralization of power and cooperative behavior—shared sacrifice—based on common interests should be the tools used to achieve the goals of sustainability.
- Coordinated and interdisciplinary approaches should be adopted to reach farmers and other relevant groups. A host of different, uncoordinated activities by government departments often result in waste of resources and confuses the farmers, even harming their interests. The team approach, which includes farmers from the word go, should promote measures that have been tested on a small

scale first. The blue-print approach, often used by government departments and agencies, should be replaced by a trial-and-error method that is capable of correcting itself, when experts and/or farmers so advise.

Below we list further measures we would like to see adopted.

- In the research and extension systems: communication of a central message to workers and farmers—that a *scientific* approach should be an essential part of daily life. The effectiveness of the message would depend on the level of education of farmers and the credibility of the agents of change.
- Farmers' participation in all agriculturally related activities and legislations is the the key to effective public policy.
- In communicating new technologies to farmers—technology transfer—the first test should be with "successful" farmers. They, in turn, will act as agents of change in their communities. This technology transfer process should include:
  1. Coordination and support for local agricultural research. In doing this, universities and research institutes should emphasize the sustainable use of natural resources and minimization of pollution;
  2. Analysis of production constraints. Measures should be adopted to alleviate them by a process of continuous monitoring and evaluation by the research and extension service;
  3. Setting the focus on reducing the gap between the high yield-levels and the actual farm average;
  4. Developing crop technology with optimum use of chemicals; to breed and adopt early maturing, resistant varieties (including hybrids) with high yields; and to promote leguminous crops in rotations and intercropping.
- The use of input subsidies should be minimized, unless they are essential for promoting activities which contribute directly to the reduction of environmental pollution and resource degradation.
- Government agents should be expected to play an important and effective role in distributing information. Good information adds to the efficiency of everyone involved in promoting sustainable agriculture. This applies both to the quality of data in research and extension services and to the information available to farmers on production inputs and technologies, postharvest management technologies, and market conditions, including price trends and stocks.
- Population growth and control: Our experience with population trends is that the process of development—focusing on growth of

incomes and opportunities for the poor—acts as a brake on population growth. Public policy on family planning, however energetic and resourceful, works best when the development process directly and positively affects the status of the poor. The cause of sustainability is best served by emphasizing population control as an integral part of the development process.

Part 2

Environmental Aspects

4

Environmental Sustainability and the Redesign of Agroecosystems

Stuart B. Hill

At some point in the future, agronomists will wonder how presentday scientists could have continued to knowingly advocate the expansion of specialized production systems that emphasize practices such as chemically managed row-crop monocultures, which result in soil erosion and degradation, and water exhaustion and contamination; or knowingly to stand idly by in the face of deforestation and desertification, loss of biodiversity, displacement of farmers and loss of rural communities, and increased dependence on nonrenewable resources, synthetic chemicals and antibiotics, subsidies, and markets that meet distant luxury desires versus local basic needs (Brown et al. 1984).

My explanation of why this madness continues may be as shocking for the readers as witnessing the present level of degradation is for me. Throughout history we have invariably blamed others for our tragedies—the gods, other nations, certain groups within society, lack of resources and power, multinationals, and political incompetence—but we have rarely examined the contribution of our own behavior, accepted our responsibilities, and set out to change our behavior. Thus my analysis of the situation is primarily psychosocial, rather than political, and that is exactly what makes such a proposition so difficult to accept, because for me this requires that I first recognize and act on my responsibilities and change myself before pointing fingers at others.

My thesis is that, because throughout history people have been psychologically wounded (Demause 1982), they have frequently established inappropriate goals, and have repeatedly done inappropriate (and unsustainable) things to achieve them, things that continue to cause major harm to both people and the planet (Hill 1992; Meadows et al. 1992). For example, the widespread hunger for power and control within families and societies, and the common obsession with the elimination of enemies, including insect pests, may have their roots in the control and manipulation of children and their treatment in some families as "pests." As

47

societies we act out these distresses with devastating consequences that range from widespread malnourishment and starvation to mental depression and degenerative diseases, to environmental degradation, to social decay, to economic crises and political unrest.

If we were managing our lives and nations in appropriate ways we would not be in this mess. Yet the predominant responses to this "feedback" from our behavior have been denial, conducting endless studies of each problem in isolation, the seeking of curative (magic-bullet) solutions, and making symbolic versus substantive changes to the design and management of our affairs. This not only reduces the options available to future generations, but perpetuates the problems and deprives each of us of experiencing a sense of meaning and fulfillment within our lives. In this paper I will argue for environmental sustainability, discuss what this implies for the design of national and local food systems, and suggest what must be done to implement such sustainable systems.

Definition of Sustainability

The foregoing observations both raise, and provide possible answers to, a number of questions. Why, for example, do we not yet have a logical and universally acceptable definition of sustainability, and why have most institutions emphasized economic over environmental sustainability (WCED 1987)? Planet Earth is our environment and our home, and "absolute" requirements (for water, air, nutrients, freedom from biocides, etc.) must be met if it is to remain a place in which present and future generations can survive. Economics, on the other hand, should be used as a tool to help us to live in ways that are consistent with our higher values, and as a consequence its characteristics are "relative." When used appropriately, economics is a tool for evaluating the costs and benefits associated with alternative courses of action. To achieve sustainability we must learn to conduct our affairs within the limits of environmental absolutes, and not continue to delude ourselves that we can only do this if we can afford it (Hill 1990a, 1991).

The widespread failure of our societies to institutionalize environmental ethics is not surprising; unlimited exponential growth, rather than restriction, has characterized human behavior in the industrialized world, particularly since the middle of the last century. It has been estimated that at that time 94 percent of the energy used in the world came from the muscles of people and domestic animals; now over 94 percent comes from fossil fuels (Othmer 1970). In fact, industrialized societies have functioned very much like drug addicts: willing to do almost anything to ensure ongoing access to the desired resources, including going to war to secure this access, and blind to the consequences of the uses of these resources and of our addictive dependence upon them (Slater 1980; Schoef 1987). It follows from this that a genuine definition of sustainability must necessarily spell out restrictions on our behavior and provide guidelines for appropriate goals.

Environmental sustainability implies the following:

1. Meeting the basic needs of all peoples, and giving this priority over meeting the greeds of a few
2. Keeping population densities, if possible, below the carrying capacity of the region
3. Adjusting consumption patterns and the design and management of systems to permit the renewal of renewable resources
4. Conserving, recycling, and establishing priorities for the use of nonrenewable resources
5. Keeping environmental impact below the level required to allow the systems affected to recover and continue to evolve

An environmentally sustainable agriculture is one that is compatible with and supportive of the above criteria.

Deep Sustainability and Resource Maintenance

A more important question relates to why these directives have yet to be taken seriously by most individuals and societies. To help recognize these real issues I distinguish between *shallow* (short-term, symbolic) and *deep* (long-term, fundamental) sustainability (Table 4.1). Shallow sustainability focuses on efficiency and substitution strategies with respect to the use of resources. It usually accepts the predominant goals within society without question, and aims to solve problems by means of curative solutions. Deep sustainability, in contrast, re-evaluates goals in relation to higher values and redesigns the systems involved in achieving these goals so that this can be done within ecological limits (Hill 1991). This approach focuses on solving problems by prevention, by creating healthy environments.

This distinction is particularly evident within the food system, and is especially clear with respect to our approaches to pest control (Hill 1984, 1990b). Thus, conventional agriculture's dependence on pesticides exemplifies a curative approach that has numerous negative side effects on both people and the environment. Although the efficient use of pesticides and the substitution of biological controls reduces these side effects, neither of these shallow approaches confronts the causes of the problem—the design and management of the agroecosystem—which must be changed if permanent deep solutions are to be found. To protect agroecosystems (and indeed nations) from problems, and to increase the resilience of these systems, efforts must be made to build up and maintain their ecological

Table 4.1: Comparison of Three Approaches to Sustainable Agriculture

Unsustainable	Shallow Sustainability		Deep Sustainability
Conventional	Efficiency	Substitution	Redesign
Examples			
Factory farm	Low-Input and Resource Efficient Agriculture	Eco-Agriculture	Permaculture, Natural and Ecological Farming
Approaches			
High power	Conservation	Conservation	Low power
Physico-chemical (soluble fertilizer, pesticides, biotechnology)	Physical/chemical/ biological (slow release, band)	Biologicals and natural materials	Bio-ecological
Imported input-intensive	Efficient use	Alternative inputs	Knowledge/skill intensive
Narrow focus, farm as factory (linear design and management)	Efficient factory	Softer factory	Broad focus, farm as ecosystem (integrated design and management)
Problems as enemies to eliminate and control directly with products and devices	Efficient control (monitor pest, Integrated Pest Management)	Biocontrols	Prevention, selective and ecological controls (pests as indicators)
Goals			
Maximize production (neglects maintenance)	Maintain production while improving maintenance	Improved maintenance	Optimize production (emphasizes maintenance)
Create demand, manipulate wants			Meet real needs

integrity or natural capital. In most countries, farmers are rewarded for productivity, but not for this kind of rehabilitation and maintenance, and indeed they are often penalized economically for spending time and energy on the latter. Until ways are found to support farmers who spend time designing their agroecosystems to conserve such resources as water, soil, and natural pest control agents, we can expect the continued erosion of these and other resources through neglect and degradation.

Psychological Roots of Shallow Sustainability

It is tempting to assume that our neglect of maintenance is determined only by economics. This is reflected in the common complaint of farmers that "I wish I could afford to farm in the way I know how." I believe, however, that this neglect has deeper psychological roots. Collectively, most of us vote for, or tolerate, governments and technologies that promise more for less; that aim to solve social problems by increasing production, export, and consumption; that ignore long-term and distant effects of current policies; and that maintain the illusion of the good life around the corner. This is, of course, blatantly irresponsible and untruthful by omission. Such formulae dig us deeper into personal and institutional pits out of which it becomes increasingly difficult to climb. We go along with, and are attracted to, such formulae because they appeal to the needy, wounded parts of our psyche. They provide what appear to be simple answers to problems that we perceive to be so complex, and so difficult and frightening to consider in their entirety (and with honesty), that we are naturally attracted to magic-bullet solutions.

Since our goals determine our actions and our actions determine the outcomes, it is essential that our goals be examined first. Because goals such as productivity, profit, and power will always lead to the treadmill of unlimited growth, and its consequential exhaustion of resources and associated impact on person and planet, they must be rejected as being incompatible with sustainability (Table 4.2). Rather we must look to what positive aims lie behind these unsustainable goals, and reformulate them as our true or higher goals. We usually say that we want to increase productivity to enable everyone to have access to the required foods, and for producers to receive a fair return, to allow them to be free to live meaningful and fulfilling lives. Thus, nourishment to support optimal physical, mental, emotional, and spiritual development, and not simply productivity, must be a primary goal. This has broad qualitative and quantitative implications for the selection of crops, the design and management of agroecosystems, and for the nature of the required institutional supports. Other goals, in addition to nourishment, that can be examined in the same way include social justice, humane treatment of livestock, and environmental and agroecosystem sustainability.

Table 4.2   Implications for the Food System of Having Sustainable Versus Unsustainable Goals

Lower Goals		Higher Goals
Productivity		Nourishment
Profit		Human Development
Power		Justice
Competitiveness		Humaneness
Growth		Sustainability
Exploitation	*Resources*	Conservation
Market Forces		Priorities
Nonrenewables		Solar/Renewables
Imported		Local
Specialized	*Farms*	Diversified
Separated		Integrated
Larger		Smaller
Dependent		Self-Reliant
Disseminator	*Extension*	Facilitator
One-Way		Two-Way
Products		Service, Skills
Isolated		Participatory
Res. Stn. Plots	*Research*	On-Farm
Short-Term		Long-Term
Single Discipline		Holistic
Technologies		Knowledge, Skills
Reactive	*Programs & Policies*	Proactive
Status Quo		Evolutionary
Defensive		Visionary
Competitiveness		Efficiency/Substitution/
Subsidies		Redesign
		Supports/Rewards/
		Penalties

Agroecosystem Design and Management

To achieve the above goals, we first need to identify, or assemble, build up, and maintain the necessary physicochemical, bioecological, and sociocultural resources. Mechanisms range from the protection of Vavilov Centers and gene pools, to soil and water conservation strategies, to collection and preservation of indigenous wisdom and skills. For example, programs for capturing water, slowing its movement across the land, and directing it to points of need, through the use of dams, retention and interception banks, swales, and channels are of paramount importance. The optimal design and siting of these structures require a profound knowledge of ecological processes, a detailed local knowledge of seasonal and topographical characteristics, and an understanding of the sociocultural constraints and opportunities (Mollison 1988; Mollison and Slay 1991).

Soil is predominantly a medium in which decomposition of the organic matter is the primary activity going on. Deprived of organic matter, soil degrades and is lost to erosion; conversely, when supplied with organic matter, the soil increases in both fertility and productivity (Hill 1989). Consequently, ways must be found to properly manage the soil, and to fix carbon, return it to the soil, and conserve it. In addition to the conservation of energy, the fixation of carbon and its build-up as organic matter in the soil is the primary strategy available to us to prevent the accumulation of carbon dioxide in the atmosphere and associated global warming. Similarly, ways must be found to support a managed succession (mimicking a natural succession) from less productive, low-demand, hardy plant covers to mixtures, including more productive and demanding crops—the latter often arranged in multistory polycultures (Mollison 1988; Mollison and Slay 1991).

With respect to our present high-input, dependent, and environmentally impacting agroecosystems, strategies must be found to support their evolution through the efficiency and substitution stages to redesigned sustainable systems (Hill 1985; MacRae et al. 1990a and 1990b). Thus, integrated pesticide and fertilizer management would give way to dependence on biological and alternative inputs, and eventually to in situ cultural methods of pest control and soil fertility maintenance (Gershuny and Smillie 1986). Special methods will be required for the rehabilitation of *agricologenic* areas, such as those degraded by salinization or contamination with toxic chemicals (Hodges and Scofield 1983). This might involve the planting and harvesting of halophytic plants and the inoculation of the soil with specialist decomposer microorganisms.

With respect to livestock, the management of wild game and crosses with compatible domestic varieties should be pursued because of the higher ecological efficiency, and higher nutrient density of the meat, associated with the former (Renecker and Hudson 1991). Matching herd size to carrying capacity and the design of appropriate systems of rotational grazing that take into account the brittleness of the pasture area are of particular importance in the achievement of range sustainability (Savory 1988).

Regulation, Monitoring, and Control

Ecological systems are being destroyed throughout the world partly because the marketplace rewards short-term productivity and not long-term resource maintenance. Consequently, governments have a particular responsibility to compensate for this market deficiency, and must resist or redesign such externally imposed controls, such as the General Agreement on Tariffs and Trade (GATT), which by reinforcing the primacy of the market and of international trade, weakens the ability of a nation to practice sustainable resource management and to protect its environment (Shrybman 1992). Specific instruments include supports, rewards, and penalties (Hill 1982):

In the longer term, however, a broader approach is required—one that aims to remove the main barriers to the achievement of a sustainable agriculture. This will include improving access to appropriate information, resources, and technologies, development of new skills, provision of a broad range of institutional supports (MacRae et al. 1990b), development of sustainable visions, higher levels of awareness, and the empowerment required to take the necessary action (Hill 1990a, 1991). Alternative visions for problem solving (MacRae et al. 1989) and for farms and farmers are listed in Tables 4.3, 4.4, and 4.5.

Achieving such visions will, however, eventually require that we examine the psychological roots of our current behavior, and of the widespread lack of vision, awareness, and empowerment, and to find ways to remedy the situation. Alternative visions are still at an early stage of development. With respect to the development of visions that could have relevance to Egypt, the work of Altieri (1987), Mollison (1988), and Savory (1988), the proceedings of an IFOAM conference in Burkina Fasso (Djigma et al. 1990), and the journal *ILEIA,* deserve special attention. Within Egypt, the work of Ibraheim Abouleish is exemplary.

In the final analysis, it will be the awareness, empowerment, vision, and values of people, and their appropriate use of science and technology, that will enable us to achieve sustainable food systems. We need repeatedly to remind ourselves that however powerful our science and technology might be, alone they will not be able to achieve sustainability; this will be realized only through our own psychosocial evolution (Hill 1990a, 1991). Indeed, the ability of a single, aware, and empowered individual to bring about meaningful change should not be underestimated. Assume you are that individual!

Table 4.3 Alternative Problem-Solving Paradigms

Reductionist, Technocentric	Holistic, Ecocentric
Cure symptoms (eliminate enemies)	Prevent, respond to multiple causes, stresses (problems regarded as indicators)
Inputs (disrupt self-regulatory mechanisms, temporary solutions)	Design and management, benign inputs (self-maintaining/regulating systems supported by cultural practices, permanent solutions)
Single, simple, direct, instant, narrow focus (magic-bullet, single discipline)	Multifaceted, complex, indirect, long-term, broad focus (multi/trans-disciplinary), decentralized
High-power, physicochemical (synthetic), imported product, expert (high-risk, expensive, dependent), centralized control	Low-power, bio-ecological, on-farm/local (low-risk, inexpensive, independent), decentralized
Technology intensive	Knowledge/skill intensive, selected technologies
Inflexible, ignores freedom of choice, disempowering	Flexible, respects freedom of choice, empowering
Temporary solutions	Permanent solutions
Unexpected disbenefits (environment, health, etc.)	Unexpected benefits
Incompatible with higher values	Compatible with higher values

Table 4.4 Some Characteristics of Farms Before and After Transition to Sustainable Agriculture

Before	After
Bare soil	Cover crops, intercrops, nonrow crops, mulches
Monoculture or row-crop rotations	Rotations including soil-improving crops
Unmanaged field borders	Windbreaks and insectary plants
Exported nutrients replaced by synthetic soluble inputs	Recycling, soil formation, and N_2 fixation
Manure — Waste disposal	Optimal management, composting
Pesticides and antibiotics (curative solutions)	Cultural and biological controls (preventative)
Grain/concentrate feeds	Forage included
Large, expensive, unmodified machinery	Smaller, modified, appropriate
Fossil-fuel based	Solar and renewable
Specialized production and marketing	Diversified

Table 4.5   Some Characteristics of Farmers Before and After Transition to Sustainable Agriculture

Before	After
Role	
Recipients of information	Exchangers, generators
Technology users	Developers
Exploiters of resources	Stewards (maintenance)
Economic viability emphasis	Nourishment of people
Approach	
Waiting for help	Experimenting
Seeking quick-fix (curative) solutions	Understanding causes
Competing	Collaborating
Controlling inputs and processes	Designing and managing
Specialization	Diversification
World View	
Dependence (rights)	Self-reliance (responsibilities)
Helplessness	Empowerment
Enemy-oriented (identify, attack, eliminate)	Collaboration, indicators (identify, facilitate, respond)
Problem control	Health promotion

References

Altieri, M.A. *Agroecology: The Scientific Basis of Alternative Agriculture*. 2nd Ed. Boulder, Colorado: Westview Press, 1987.

Brown, L.R. et al. *State of the World: A Worldwatch Institute Report on Progress Toward a Sustainable Society*. New York: W.W. Norton, 1984.

Demause, L. *Foundations of Psychohistory*. New York: Creature Roots, 1982.

Djigma, A., E. Nikiema, D. Lairon, and P. Ott (editors). *Agricultural Alternatives and Nutritional Self-sufficiency for a Sustainable Agricultural System that Respects Man and His Environment*. Proceedings of the IFOAM Seventh International Scientific Conference, Ouagodoujou, 1989. Tholey-Theley, Germany: IFOAM, 1990.

Gershuny G., and J. Smillie. *The Soul of Soil: A Guide to Ecological Soil Management*. 2nd Ed. Weedon, Québec: Gaia Service, 1986.

Hill, S.B. "A Global Food and Agriculture Policy for Western Countries: Laying the Foundations," *Nutritional Health*, 1982, 1(2): 108–117.

———. "Controlling Pests Ecologically," *Soils Association Quarterly Review*, 1984, 13–15.

———. "Redesigning the Food System for Sustainability," *Alternatives* 1985, 12(3/4): 32–36.

————. "The World Under Our Feet," *Seasons,* Winter 1989, 15–19.

————. "Ecological and Psychological Prerequisites for the Establishment of Sustainable Prairie Agricultural Communities," in *Alternative Futures for Prairie Agricultural Communities.* Edited by J. Martin, Edmonton, Alberta: University of Alberta, 1990a.

————. "Pest Control in Sustainable Agriculture," *Proceedings of Entomological Society of Ontario,* 1990b, 121: 5–12.

————. "Ecovalues, Ecovision, Ecoaction: The Healing and Evolution of Person and Planet," *Absolute Values and the Reassessment of the Contemporary World.* Vol. 2. New York: ICF Press, 1991, 1019–1033.

————. "Changing Ourselves to Change the World," *Manna,* Newsletter of the International Alliance for Sustainable Agriculture. 1992, 8(4): 1–5.

Hodges, R.D., and A.M. Scofield. "Agricologenic Disease: A Review of the Negative Aspects of Agricultural Systems," *Biol. Agric. Hort.* 1983, 1: 269–325.

MacRae, R.J., S.B. Hill, J. Henning, and G.R. Mehuys. "Agricultural Science and Sustainable Agriculture: A Review of the Existing Scientific Barriers to Sustainable Food Production and Potential Solutions," *Biol. Agric. Hort.* 1989, 6(3): 173–219.

————. "Farm-scale Agronomic and Economic Conversion From Conventional to Sustainable Agriculture," *Advanced Agronomy.* 1990a, 43: 155–198.

MacRae, R.J., S.B. Hill, J. Henning, and A.J. Bentley. "Policies, Programs and Regulations to Support the Transition to Sustainable Agriculture in Canada," *American Journal of Alternative Agriculture,* 1990b, 5(2): 76–92.

Meadows, D., D. Meadows, and J. Randers. *Beyond the Limits.* New York: Chelsea Green Press, 1992.

Mollison, B. *Permaculture: A Designer's Manual.* Tyalgum, Australia: Tagari, 1988.

Mollison, B., and R.M. Slay. *Introduction to Permaculture.* Tyalgum, Australia: Tagari, 1991.

Othmer, D.F. *Man Versus Materials.* Transactions of New York Academy of Sciences, Series II, 1970, 32(3): 287–303.

Renecker, L.A., and R.J. Hudson (editors). *Wildlife Production: Conservation and Sustainable Development.* Fairbanks, Alaska: University of Alaska Press, 1991.

Savory, A. *Holistic Resource Management.* Washington, D.C.: Island Press, 1988.

Schoef, A.W. *When Society Becomes an Addict.* San Francisco: Harper and Row, 1987.

Slater, P. *Wealth Addiction.* New York: Dalton, 1980.

World Commission on Environment and Development. *Our Common Future.* New York: Oxford University Press, 1987.

5

Environmental Sustainability of Egyptian Agriculture: Problems and Perspectives

Asit K. Biswas

Egypt, said Herodotus, is the gift of the Nile. More than two millennia after the Greek historian's visit, and in spite of extensive technological developments, this still is not an overstatement. Life in Egypt would be impossible without the waters of the Nile. Napoleon Bonaparte reconfirmed the observation of Herodotus during the French occupation of Egypt in the late eighteenth century, saying: "If I was to rule a country like Egypt, not even a single drop of water would be allowed to flow to the Mediterranean Sea." While Napoleon was not aware of the importance of allowing some discharge of the Nile waters to the sea for salt balance and other environmental reasons, the general direction of his thinking was undoubtedly correct.

In Egypt, the major constraint to agricultural development is water and not land. Even after the construction of the High Aswan Dam (HAD), which radically altered the water-use patterns in Egypt, only about 7.49 million feddans, which is less than 4 percent of the country's land area, is cultivated at present (Biswas 1991a). Clearly it is impossible to raise and meaningfully discuss all the agricultural sustainability issues for Egypt in one chapter. However, since most of the papers at the conference were from the agricultural professions, the main thrust of this chapter will be on the role of water for irrigation.

Sustainable Development: Concept and Issues

The concept of sustainable development is not new: the general philosophy behind it has been recognized for many centuries. However, there is no agreed definition of sustainable development. Currently more than a hundred definitions exist, and they could differ in significant ways. Such definitions are somewhat simplistic, internally inconsistent, and vague for actual use in policy formulation, as well as for planning and implementa-

tion of specific projects, except in a very general sense. And any development process that does not consider the achievement of a reasonable and
equitably distributed level of economic well-being—that can be maintained for many generations—cannot be sustainable. This fact is not
properly reflected in many definitions.

In spite of present rhetoric, operationally it is still not possible to
identify a development process that can be planned and then implemented
and which would be inherently sustainable, however it may be defined.
We have had more success in identifying aspects of development which
are *un*sustainable (and then taking remedial steps to reduce or eliminate
those undesirable effects) than in devising a holistic process that is intrinsically sustainable from the beginning (Biswas 1991b).

Many issues are important for sustainable agricultural development,
but three aspects are particularly worth noting from a policy point of view.

1. Short-term versus long-term considerations: The concept of sustainable
development assumes that the process selected would be viable over the
long-term—though the issue of what constitutes long-term has neither
been clarified nor featured much in current discussions. The time factor,
either inadvertently or because of its complexity, has been basically left
fuzzy. For example, does sustainability cover fifty years, or one hundred,
five hundred, one thousand years, or even longer? Some have spoken
vaguely of "several" generations.

To consider even the lower figure of fifty years, there is a fundamental
dichotomy as to its use in the real world. For example, even in the newly
reclaimed areas of Egypt, the time horizon of farmers would extend to
from two to three years; certainly no more than five years. The overriding
objective of nearly all farmers is to maximize economic returns from their
agricultural activities within this time horizon. The mindset is based on
maximizing profits over a continual series of short periods. If the short-
term benefits could have long-term costs, even to themselves (e.g., in terms
of soil erosion, salinity, etc.), generally, short-term considerations have
won over the long-term implications. Accordingly, even if the societal goal
is to achieve (long-term) sustainable development, in reality the main
objective of a vast majority of farmers extends normally to short-term
survival. Any plan which does not specifically consider this fundamental
conflict, and then attempts to identify realistic alternatives to overcome
the problem, is doomed to fail.

2. Externalities: Externalities occur when private net benefits do not equal
social net benefits. Farmers and large agricultural estates operate primarily on the basis of their own private costs and benefits. If they perceive
opportunities which could reduce their costs and/or increase benefits, they
often take actions which could be beneficial to them but are unlikely to

serve the common good. Commonplace examples include use of excessive irrigation water by farmers in the headreaches of canals, which means the tailenders have insufficient and/or unreliable water supply. This, in turn, could decrease the crop yields and thus the incomes of tailenders substantially. Similarly, wastes from agroprocessing industry are discharged to canals and rivers, which can impair existing water-uses downstream.

Such costs could be internalized, at least conceptually, through taxes, subsidies, and regulations. But in reality, even in developed countries, it has not been easy to internalize the externalities for four important reasons. (1) The calculation of precise value of externalities is very difficult. (2) Frequently, politically powerful individuals and organizations vociferously defend their considerable private advantages against a large number of unorganized and disadvantaged individuals (who may be experiencing additional costs indirectly). (3) Externalities can develop over time, and thus there can be a time gap before those affected realize the real costs. (4) Regulations to control such externalities in nearly all developing countries have proved ineffective and expensive.

3. Risks and uncertainties: A major issue confronting sustainable agricultural development has to do with risks and uncertainties associated with complex systems. For example, with increasing population in Egypt, there is no question that land and water have to be intensively used to maximize production. The fundamental question for which there is no real answer is, Up to what level can Egypt's agricultural production system be intensified without sacrificing sustainability? What early warnings could indicate the beginning of a transition process from sustainable to unsustainable? What parameters need to be monitored to indicate that such a transition is about to occur—or even is occurring?

Our present knowledge is inadequate even to identify the parameters that could indicate the transition. We cannot accurately detect, much less predict, the transition from sustainable to unsustainable. And agricultural systems are variable by nature. Their fluctuations can be so great that statistically significant data would be very expensive or even impossible to obtain in order to state categorically whether such variations are natural or signs of unsustainability. Additional factors like potential climatic changes increase the uncertainty to a manifold degree (Abu-Zeid and Biswas 1992). One is then confronted with the difficult issue of being able even to identify the direction of change, let alone the degree of change.

Fundamental issues of this type must be successfully resolved before the concepts of sustainable agricultural development can be holistically conceived and implemented. Much lip service is given to sustainable agricultural development at present, but most of the published works on this subject are either general or continuations of "business as usual" that have only been given the latest trendy label, "sustainable development."

If sustainable agricultural development is to become a reality, national and international organizations would have to address the real questions. So far they have not done so in any measurable and meaningful fashion. Unless the rhetoric can be translated into reality, sustainable development will remain a trendy catchword for a few years, and then fade away like that earlier concept, ecodevelopment.

Land Resources

For an arid country like Egypt, the prime factor which makes land productive is water, and an analysis of arable land can best be divided as the pre- and post-HAD periods. Fortified by the increased and more reliable water supply that was made possible by the construction of this dam, and assisted by other technological developments, it has been possible both to intensify cultivation in the old lands and to expand agricultural activities in the new lands. Construction of the HAD confirmed the fact that the supply of arable land in Egypt was not necessarily inelastic, as had been assumed for centuries. But the availability of arable land per capita has declined from about 0.50 feddans in 1897 to 0.13 feddans in 1990, due to the rapid increase in population—an increase from about eleven million to fifty-five million in the same period. Note also that the rate of increase of total arable land during the 1980–1990 period (addition of about one million feddans) is highly unlikely to be maintained in the future.

Land reclamation is dealt with in other chapters, but not much has been written on the *loss* of productive land in Egypt, except in a general and anecdotal fashion. Estimates of land-loss available at present range from a low of 20,000 feddans (Parker and Coyle 1981) to a high of more than 100,000 feddans (World Bank 1990) per year. All these estimates are based on anecdotal observations only. The problem of calculating land-loss is compounded by four factors:

1. The net area of cultivated land can only be guessed at; the last agricultural census was in 1961.
2. Land reclamation statistics refer only to gross areas: reliable data are not available on areas that are not fully reclaimed or are unproductive and/or abandoned.
3. Information is not available on land-losses due to urbanization, even for very specific years.
4. Current estimates of land-loss due to waterlogging and salinity are so vague as to be meaningless.

The environmental literature on Egypt is full of anecdotal or superficial estimates on land-loss and these have been masquerading as realistic data (World Bank 1990; Haas 1990; Kishk 1986). My best "guestimate" of the current annual loss of agricultural land due to urbanization would be of the order of 30,000 feddans. Assuming this estimate to be reasonable, this means that land reclamation efforts have increased the agricultural land base but very modestly. Egypt must give urgent attention to reducing the loss of arable land to urbanization—first because with continually increasing population existing agricultural land should not be lost; second because land reclamation is an expensive process, it is desirable not to lose land areas that are already productive and then try to compensate by reclamation; and third because land lost due to urbanization is often more productive than the reclaimed land.

Water Resources

In spite of the critical importance of water to Egypt, unfortunately no reliable national estimates of water balance are available. Neither are reliable estimates of current water requirements for various sectors like agriculture, industry, municipal, navigation, and hydropower generation. Without such estimates, rational planning for agricultural water availability and its efficient use for the medium- to long-term is not an easy task.

The agricultural sector is by far the major user of water, and we should have a clear idea of the quantity of water that is likely to be available to this sector in the future, and the modes of its utilization. The 1990 estimates from the Ministry of Public Works and Water Resources (MPWWR) indicate that agriculture accounts for 49.7 billion m^3 per year, which is 84 percent of the total water-use in the country. This amount does not include an annual estimated loss of 2 billion m^3 from canals, main, lateral, and sublateral, mainly due to evaporation. Total annual evapotranspiration losses in Egypt are estimated at 34.8 billion m^3, which accounts for the bulk of agricultural water-use. This figure, however, is misleading. Really it is the difference between releases at Aswan and other outflows and usages, so this estimate includes not only crop-used water but also "unaccounted for" water-use (Biswas 1991a).

Water Availability for Agriculture

The downward trend in the amount of water available for agriculture is likely to accelerate in the post-2000 period, primarily because of sharply increasing demand from the industrial sector. The MPWWR estimates of industrial water-use in 1990, based on extrapolation of a very limited field survey in 1980, was 4.7 billion m^3. It is expected to increase to 7.6 billion m^3 by the year 2000. In the absence of reliable studies on industrial water-use in Egypt, any forecast can only be considered very preliminary, even for the projection to the year 2000. This estimate is likely to be adjusted significantly as more reliable data are available.

Taking studies in other developing countries as an indication, the general trend, however, is likely to be accurate (Biswas 1991a). Water requirements for the domestic sector will continue to rise—because of increasing population and higher demand from people who attain a better standard of living. Industrial demand would also increase very significantly. Since domestic and industrial demands would continue to receive higher priority, the share of water for agriculture would decline. Hence, any sustainable agricultural strategy for Egypt must be based on a declining share of available water.

Role of Water Conservation

Since the agricultural sector will have to get used to the concept of doing more with less—its reduced share of future water—water conservation must be a priority item in any agricultural strategy. Major policy options are available to increase water-use efficiency: water-pricing, improved operation, maintenance and rehabilitation of irrigation systems, reduction of all types of water losses, development of water-conserving varieties of crops, substitution of water-intensive crops by those using less per unit of output. They all should be pursued vigorously. The agricultural sector, as the major user of water, naturally has the highest potential for water saving.

Groundwater

Serious and extensive studies of groundwater availability are of comparatively recent origin in Egypt and, as a general rule, information on groundwater is less reliable than that on surface water. Current estimates indicate that the total groundwater available in the Nile Valley and the Delta is about 500 billion m^3. The existing annual rate of extraction in this region for domestic, industrial, and agricultural purposes is estimated at 2.6 billion m^3. This can probably be increased on a sustainable basis to about 4.9 billion m^3, which is estimated to be equivalent to the annual recharge rate (Biswas 1991a).

The policy issues are more complex for groundwater use in the Western Desert, New Valley, and Sinai, since this is fossil water and thus not a renewable resource. For example, preliminary estimates indicate that the total groundwater storage in the New Valley is of the order of 40,000 billion m^3, with salinity levels varying between 200 and 700 ppm. Use of this or any other fossil water would depend on its quality, cost of pumping, and the probable economic return over a fixed period. Consideration has also to be given to the potential socioeconomic implications for the area when water can no longer be drawn economically. While the use of fossil groundwater has already started in Egypt, there is no clearly enunciated policy at present. It is essential that a clear government policy be developed.

Treated Wastewater

Reliable estimates of the amount of treated wastewater likely to be available by the year 2000 and in succeeding years are not available, but preliminary estimates can be made on the basis of the current plan for wastewater treatment for the Greater Cairo area. Assuming an average wastewater production of 340 liters per capita per day, the total amount of wastewater from the Greater Cairo area could be of the order of 1.9 billion m^3 annually by the year 2010. Treated wastewater could thus become an important new source of water, and this factor should be properly considered in any new agricultural development strategy.

For the environment and the country's resource base, treated wastewater has some benefits as well as constraints. On the benefit side, treated wastewater contains fertilizers—nitrogen (N), phosphorus (P), potassium (K)—and some micronutrients—e.g., metals. Hence, in terms of agricultural productivity, treated wastewater is normally more beneficial than irrigation water. However, it does not contain the required major nutrients in optimal proportions for crop growth, nor in large amounts (Biswas and Arar 1988). In addition, on the basis of limited data available, it appears that wastewater contains less N, P, and K in Egypt than in developed countries. While precise reasons for this anomaly cannot be given at present, the P content could be less due to comparatively low use of synthetic detergents in Egypt.

As to constraints, use of treated wastewater could pose environmental and health problems and it is necessary to ensure that all treatment plants function properly, and that appropriate monitoring is done regularly. Experience in the use of treated wastewater in Egypt is limited. Some time ago, an interdepartmental committee was established on this subject under the leadership of the Academy of Scientific Research and Technology, but the committee has made little progress. It is, however, essential that some major pilot irrigation projects be established when treated wastewater becomes available from the Greater Cairo project.

Water Quality Issues

It is not enough to consider only issues related to water quantity without any reference to its quality. Water quality has major implications in terms of agricultural water-use, and agricultural activities in turn could affect water quality, which could reduce water availability for other sectors. Unfortunately, with the data now available, no reliable statement can be made on the status of water quality in Egypt—a situation similar to that in other developing countries. On the basis of limited data and some anecdotal information, it is clear that water pollution is a serious problem in many canals and drains, as well as for parts of the Nile. There are fundamental problems that need to be addressed promptly, among them the following:

1. There is no national strategy on proper collection, analysis, review, and dissemination of water quality data. Nor is there a rational, comprehensive water quality management plan that can be implemented, although efforts are now being made to address these issues.

2. The important issue of the institutional arrangements for water quality data collection and data management plans have received scant attention. Within MPWWR, water quality data are being collected primarily by three institutes under the Water Research Center (WRC)—the Nile Research Institute (NRC), the Drainage Research Institute (DRI), and the Research Institute on Groundwater (RIGW), each with its respective area of interest. Realistically, in the long term, MPWWR would have to develop a functional division responsible for collection of routine water quality data. The attempt must also be made to make the existing water quality data management systems compatible. At present, the hardware and software systems used by NRI, DRI, and RIGW are not compatible. A clear picture of the country's water pollution problems could not be obtained even if the data were available, or not without significant additional work. The data collection systems have been initiated only recently: compatibility of the systems should receive urgent attention.

3. There are serious shortcomings in the present practice of taking one or two samples per year on a limited but fixed number of water quality parameters along the Nile and selected canals and drains. Even the proposed plan to extend it to four sampling runs per year would be of very limited use for proper water quality management. Any usable water quality monitoring program has to be more flexible—with regard to frequency of sampling, sites where samples are taken, and parameters sampled.

4. Much progress is needed in the area of water quality control and quality assurance. Proper training of an adequate number of water quality analysts is essential for laboratories to produce consistently reliable analyses.

Control of Sources of Water Pollution

There are three major sources of pollution in Egypt: domestic, industrial (including agroprocessing), and agricultural. Domestic (household) wastewater requires at least primary treatment, and where necessary, secondary and tertiary treatments. Without functional treatment facilities, water pollution from domestic sources cannot be controlled.

The pollution of the Nile, canals, and drains from industrial effluents is probably more complex—more complex than either domestic or agricultural pollution. The government currently owns and operates 367 industrial facilities in Egypt. Most of these are related to the agricultural sector, though individual facilities may be owned by a ministry other than the Ministry of Agriculture and Land Reclamation (MALR). The Ministry of Industry is the major owner (72 percent), followed by the ministries of economics (9 percent) and supplies (8 percent). The remaining 11 percent are owned by the ministries of defense, electricity, agriculture, housing, and health. Of these industries, some 35 percent discharge effluents directly to the Nile or through agricultural drains; 15 percent discharge to canals that are sources of water for irrigation and drinking; and the rest to coastal lakes and the Mediterranean Sea, or, through land application, to deep wells and irrigation. For the private sector, very little data are available in terms of effluent quality and quantity. The pollution potential from private industry—a rapidly growing sector—is underestimated at present.

In the absence of reliable information, little can be said authoritatively about industrial water pollution in Egypt, but the trends are clear. Industrial water pollution is a rapidly growing problem. In parts of Egypt, bodies of water are already seriously contaminated. They would violate World Health Organization (WHO) guidelines for drinking water.

The optimal solution is to control pollution at source, and for the agricultural sector consideration should be given to more efficient use of pesticides (including integrated pest management) and fertilizers.

The legal basis of controlling water quality has been Public Law 48 of 1982, "Protection of the River Nile and Waterways from Pollution (Water Act)." The law established stringent effluent standards for organic and inorganic pollutants. It also provided strict sanctions against polluters. But the law was poorly drafted, and the standards stipulated were too strict and rigid, with no possible recourse. No serious studies were carried out to determine the potential problems for implementing the law in terms of the country's economic, technological, and social conditions, e.g., availability of adequate funds, trained manpower, sophisticated laboratories, and transportation facilities for analyses, monitoring, inspection, and enforcements. Not surprisingly, the law has been basically ineffectual. Ironically, some of the public sector companies are the worst offenders. Shortly after the law was promulgated, the government was forced to grant

Environmental Aspects

dispensations to polluters because of their inability to comply with the legal requirements. Public Law 48 has to be amended significantly, and this amendment cannot be done on the basis of theoretical considerations only. Serious studies have to be carried out if the law is to have a sound techno-economic basis, improving its implementation potential.

Conclusion

In order to account for an increasing population and a higher average standard of living, Egypt must use all available land, water, and related resources intensively and efficiently. The real question is not whether these resources should be used intensively and extensively, but rather how this can be achieved in an environmentally sound fashion so that the development process can remain sustainable over a long term. No sane person would argue with this general goal. The real problem arises when decisions have to be made on how to design, implement, and maintain an intrinsically sustainable agricultural development project. Some progress has been made during the past two decades, especially on some aspects of unsustainability, but we still do not know enough on how all aspects of a project, plan, or program should be handled and coordinated so that its long-term survival can be ensured.

References

Abu-Zeid, M., and Asit K. Biswas. "Impacts of Agriculture on Water Quality." *Water International,* 1990. 15, no. 3: 160–167.
———. *Climatic Fluctuations and Water Management.* Oxford: Butterworth-Heinemann, 1992.
Biswas, Asit K. "Land and Water Management for Sustainable Agricultural Development in Egypt: Opportunities and Constraints," Report to Policy Analysis Division, FAO, Rome, 1991a.
———. "A Holistic Approach to Environmental Assessment of Water Development Projects," International Symposium on Water Resources Management at Global and Regional Scales, Otsu, Japan, 1991b.
Biswas, Asit K., and A. Arar. *Treatment and Reuse of Wastewater.* London: Butterworth, 1988.
Haas, P.M. "Towards Management of Environmental Problems in Egypt," *Environmental Conservation,* 1990. 17, no. 1: 45–50.
Kishk, M.A. "Land Degradation in the Nile Valley," *Ambio,* 1986. 15, no. 4: 228–230.
Parker, J.B., and J.R. Coyle. "Urbanization and Agricultural Policy in Egypt," FAER No. 169, Economic Research Service, U.S. Dept. of Agriculture, Washington, D.C., 1981.
World Bank. "Egypt: Environmental Issues," Draft Discussion Paper, World Bank, Washington, D.C., 1990.

Part 3

Development of Natural and Human Resources

6

Egypt's Water Resource Management and Policies

Mahmoud Abu-Zeid

Egypt's prosperity still depends largely on the agricultural sector and its productivity. However, the fearful increase in population—growing at 2.6 percent annually—represents the greatest challenge to Egypt's future.

The River Nile is the principal water resource and is expected to remain so for years to come, supplying Egypt with more than 95 percent of its present water requirements. With the limitation of Egypt's share and the complexity of developing its north-bound water inflow, the need for rationalizing the different uses of water acquires utmost importance, and to this end the Ministry of Public Works and Water Resources (MPWWR) has adopted several policies.

The High Aswan Dam (HAD) was established as a long-term storage reservoir to ensure a constant and regular inflow for both Egypt and Sudan. However, the drought period that prevailed in the region from 1979 and lasted for nine uninterrupted years has seriously affected the storage in the High Dam Lake Reservoir—a fact that motivated the country to develop various scenarios to face the probability of a recurrence of such a catastrophe. Some of those alternatives are to consider reducing, as much as possible, rice and sugarcane cultivation; to minimize the water duties for different applications; and to generate electricity during the winter closure period. The government is studying ways of storing fresh water, which is discharged to the sea during this period, and reusing it in agriculture. Linked to these proposals and plans are studies of the laws and regulations that govern water-use and coordination between ministries, organizations, and water users. A supreme ministerial committee for water has been constituted, headed by the minister of public works and water resources and including representatives from concerned agencies, to discuss water policies. This chapter deals with those policies.

Water Resources

Egypt is a very arid country. The average annual rainfall seldom exceeds 200 mm along the northern coast. The rainfall declines very rapidly from the coast to inland areas, and is negligible south of Cairo. This meager rainfall occurs in winter, in the form of scattered showers, and cannot be depended upon for extensive agricultural production. Thus, reliable availability of irrigation water is an absolute necessity for agricultural development.

Surface Water Supply

The main and almost exclusive source of surface water is the River Nile. The Nile Water Agreement of 1959 with Sudan clearly defines the division of Nile water travelling north between Egypt and the Sudan. Nearly 85 percent of the water to both countries originates in the Ethiopian Highlands. The 1959 Agreement was based on the average flow of the Nile during the 1900–1959 period. The average flow at Aswan, Egypt, during this period was 84 billion m^3. The average annual evaporation and other losses in the High Dam lake were estimated at 10 billion m^3, leaving a net usable annual flow of 74 billion m^3. Under the 1959 treaty, 55.5 billion m^3 was allocated to Egypt and 18.5 to the Sudan.

The High Aswan Dam was constructed in 1964 to assure long-term availability of water for both countries. Its lake has a live storage capacity of 130 billion m^3. The annualized Nile flow at Aswan for the past one hundred years indicates that the average flow varies slightly if shorter durations are considered. The annual discharges from the High Dam lake during the period 1968 to 1990 show that there is a potential for increasing the Nile flow at Aswan.

The Joint Egyptian-Sudanese Committee has outlined several development programs, the first of which is the construction of the Jonglei Canal. The project was expected to canalize the river channel in the Sudd region of the Sudan and thus reduce the substantial evapotranspiration losses. The construction of phase one of the canal was started in 1976 but had to be abandoned in 1983 due to security problems in southern Sudan. Initially, this phase was expected to be completed around the mid-1980s, which would have provided an additional 4 billion m^3 of water at Aswan to be shared equally by the two countries. The total loss in the Machar swamps by evapotranspiration is about 10 billion m^3. Conservation schemes in this sub-basin are expected to yield an average gain of 4.4 billion m^3 at the White Nile or about 4.0 billion m^3 at Aswan. The above estimates of water savings from the proposed conservation projects in the Upper Nile sub-basins adds to a minimum of 18 billion m^3. However, finalization of these schemes depends on agreements between the Nile basin countries and the investment requirements. A total of 7 billion m^3 was expected after the completion of phase two of the Jonglei Canal. Joint efforts are required to resume the work on the first phase of construction of the canal, in which over 70 percent of the work was completed.

The water discharge in the streams from Bahr El-Ghazal (another sub-basin of the equatorial plateau) is about 14.0 billion m^3 in a normal year, of which only 0.6 billion m^3 reaches the White Nile at Lake Noo and the rest is lost in the swamps. Proposed schemes for conserving the water of Bahr El-Ghazal are expected to yield a saving of 12 billion m^3 annually at Malakal or roughly 10 billion m^3/year at Aswan. The water lost in the Sobat and tributaries basin reaches 5 billion m^3 each year. There are no definite plans for conserving this water similar to those of Bahr El-Ghazal.

Groundwater Supply

Groundwater in Egypt can be divided into two categories. The first comprises the Nile Valley and the Delta system. The total storage capacity of the Nile Valley aquifer is about 200 billion m^3, with an average salinity of 800 ppm. Another 300 billion m^3 is the storage capacity in the Delta aquifer. The current annual rate of extraction of groundwater from the two aquifers is 2.6 billion m^3. This can be increased to a safe annual extraction rate from the aquifer system, currently estimated at 4.9 billion m^3.

Groundwater also exists in the Western Desert, generally at great depths. Most recent studies have indicated that this is not a renewable resource. Preliminary estimates indicate that the total groundwater storage in this area is of the order of 40,000 billion m^3, with salinity varying between 200 and 700 ppm. Use of this fossil water depends on the cost of pumping, depletion of storage, and potential economic return over a fixed period. Investigations in the New Valley indicate that about 1 billion m^3 of groundwater can be used annually at an economic rate. This will allow irrigation of 150,000 acres, of which 43,000 acres are already being cultivated. An additional 190,000 acres can be irrigated in the East Ewainat area (southern part of the Western Desert) by groundwater from the deep Nubian Sandstone aquifer. More studies are under way to investigate the groundwater potential within this regional aquifer. This work is being carried out in cooperation with Sudan and Libya.

Groundwater is available in Sinai in numerous aquifers of varying capacities and qualities, but it is generally believed that it is in limited quantities. Shallow aquifers in the northern coastal areas are replenished by the seasonal rainfall. The thickness of the aquifer varies between 30 and 150 m and its salinity increases from 2,000 ppm to 9,000 ppm near the coast. In the north and central parts of Sinai, groundwater aquifers are formed due to recharge by the rain storms falling and collected in the valleys. Deep aquifers with nonrenewable water exist in Sinai, where wells are drilled to a depth of 1,000 m to supply water for domestic use. The El-Arish-Rafaa coastal area in north Sinai has always been of importance. The present extraction rate from the Quaternary aquifer in El-Arish is estimated at 52,000 m^3/day. This area is now facing a state of quality deterioration in space and time. The system is being exploited and it needs to be safely managed. The groundwater investigations in South Sinai include several shallow and deep reservoirs which have a definite potential for development, but again of limited scale.

Water-Use

Total water-use in Egypt in 1990 was estimated at 59.2 billion m^3, of which 84 percent was used in agriculture. Industrial, municipal, and navigational use accounted for additional 8 percent, 5 percent, and 3 percent, respectively. Current estimates indicate that the total water-use will increase to 69.4 billion m^3 by the year 2000. The share of water-use by the agricultural and municipal sectors will remain almost similar to 1990, but the share of industry will increase by 50 percent, and the navigational use will decline very substantially.

Agricultural water-use: While the amount of water used for agriculture has declined slowly during the past decade, it still accounts for the largest share (84 percent) or 49.7 billion m^3 per year. This amount does not include an annual estimated loss of 2 billion m^3 due to evaporation from the irrigation system. Annual evapotranspiration losses are estimated at 34.8 billion m^3. The government has launched a national program for irrigation improvement and water management. Surface irrigation systems are used in most cultivated lands of the Nile Valley and Delta and their efficiency is considered to be low. Excess applications of water to crops contribute to problems of salinity and high water-tables. It must also be noted, however, that excess irrigation water contributes to groundwater, a good part of which is pumped or partially reused through cycling. All of which increases overall water-use efficiency to a reasonable level. The measured drainage water out of the system amounted to about 11 billion m^3 during 1990. In the new lands, modern irrigation systems such as drip and sprinklers are used. The government does not give permits for new water to lands unless evidence is given of the use of new irrigation technologies.

Domestic water-use: Annual domestic water-use for 1990 was estimated at 3.1 billion m^3. It is also estimated that the level of distribution losses is 50 percent. It is assumed that the domestic water-use could be held at 3.1 billion m^3 by the year 2000 by reducing distribution losses to 20 percent.

Industrial water-use: It was estimated that industry used 4.6 billion m^3 in 1990. This estimate is based on the extrapolation of the 1980 survey carried out for the Water Master Plan.

Navigational water-use: From February to September, water releases for irrigation are sufficient to maintain water levels in the Nile for navigation.

Irrigation demands from October to January, however, are not enough to maintain a navigational level in the river. This period is the peak tourist season, when tourist boats make regular sailings between Aswan and Luxor. Some 1.8 billion m^3 of water has to be released during this period to maintain the navigational level. The Esna Barrage is being rebuilt, which will provide better control of the Nile water level. By the year 2000, annual navigational water requirements could be reduced to only 0.3 billion m^3 through better control of water level and the establishment of storage in the northern lakes.

Re-use of treated water: Wastewater has been re-used indirectly in Egypt for centuries, but the first formal use of wastewater was initiated in 1915 in the eastern desert area of Jabal El-Asfar, northeast of Cairo. After primary treatment, wastewater was used for desert agriculture, bringing into cultivation an area of 2,500 acres. As new wastewater treatment plants come on-line in Cairo and other urban areas, the amounts of treated wastewater available for agricultural activities will increase steadily during the next three decades. Total wastewater available from the Greater Cairo area will increase from 0.9 billion m^3 in 1990 to 1.7 billion m^3 in the year 2000 and 1.93 billion m^3 annually by the year 2010, according to estimates.

It should be stressed that Egypt's experience with wastewater re-use is limited. From a policy viewpoint, steps should be taken to establish major pilot projects for its use in agriculture. These projects would help to convince the general population that such practices, when properly carried out, impose no risk to human and animal health.

Proper sewage treatment, in addition to providing treated wastewater, could make another major contribution to agriculture. Dried sludge can be effectively used as a soil conditioner. During 1988/89, 46,000 m^3 of dried sludge was produced and sold to farmers and other organizations at Jabal Al-Asfar and Abu-Rawash. It is estimated that at full development, the Greater Cairo wastewater project in the year 2010 would produce 3,410 tons per day of dry solids. A conservative estimate of the total annual market demand of sludge is 779,000 m^3.

Re-use of agricultural drainage water: Agricultural drainage water in Upper Egypt is discharged back into the Nile. This affects slightly the quality of the Nile water: its salinity increases from 250 ppm in Aswan to 350 ppm in Cairo. The drainage water in the Nile Delta is of lower quality, and accordingly is collected through an extensive drainage network for disposal into the Mediterranean Sea. The total amount of drainage water discharged to the sea depends on many factors, e.g., the amount of water released at Aswan, cropping patterns, and irrigation efficiency. The total amount of drainage water discharged annually has varied from 14 billion

Table 6.1   Nile Water-Flow Downstream of High Aswan Dam and Drainage Water
Flowing to the Sea

Year	Nile Water Downstream of HAD (billion m³)	Drainage Water Quantity (billion m³)	Drainage Water Salinity (mmhos/cm)
1984–85	56.40	14.12	3.71
1985–86	55.52	13.86	3.68
1986–87	55.19	13.03	3.64
1987–88	52.86	11.82	6.15
1988–89	53.24	11.12	4.63

m^3 in 1984/85 to 11 billion m^3 in 1988/89 (Table 6.1). The salinity of this water ranges between 1,000 and 7,000 ppm: 25 percent of this water in 1984 and 70 percent in 1988 had a salinity level of less than 3,000 ppm. MPWWR is imposing strict policies for the releases of Nile water downstream from the High Aswan Dam. Further decrease in drainage water quantity and increase of its salinity will occur when the irrigation efficiency is improved both in the conveying system and at the farm level.

Surveys and monitoring of the quality and quantity of the agricultural drainage water in the Nile Delta have shown that it is possible to re-use part of this water in irrigation. When salinity is low, the water is used directly. When it is high, it is mixed with fresh canal water. Water with higher salinity and water contaminated by municipal and industrial wastes cannot be used in irrigation. Under most circumstances a substantial portion of drainage water must be discharged into the sea to maintain the salt balance in the Nile Delta.

The amount of drainage water presently re-used in irrigation is 4.7 billion m^3 annually, of which 2.6 billion m^3 is in the Nile Delta, 0.95 billion m^3 in Fayoum, and 1,015 billion m^3 returned to the Nile in Upper Egypt. This is expected to be increased gradually to reach 7.0 billion m^3 by the year 2000. It should be noted that potential savings from improved water management (greater efficiency and reduced outflows to the sea as practiced in 1987/88 and 1988/89) and increased re-use of drainage water are not mutually exclusive. However, there is a real danger that salinity could increase steadily over the years. A cautious approach to increasing the use of drainage water is likely to be in Egypt's long-term interest.

Land Resources

For an arid country like Egypt, the prime factor which makes land productive is water. Thus an analysis of arable land can be best divided as

Table 6.2 Changing Patterns of Population and Arable Land in Egypt 1897–1990

Year	Population (million)	Arable Land	
		Total (million feddans)	Per Capita Land (in feddans)
1897	9.7	4.9	0.51
1907	11.2	5.4	0.48
1917	12.8	5.3	0.41
1927	14.2	5.5	0.39
1937	15.9	5.3	0.33
1947	19.0	5.8	0.31
1960	26.1	5.9	0.23
1970	33.2	6.0	0.18
1980	42.1	6.1	0.14
1990	55.0	7.2	0.13

pre– and post–High Aswan Dam periods. Fortified by increased and more reliable water that was made possible by the construction of this dam, and assisted by technological developments, it has been possible both to intensify cultivation in the old lands and to expand agricultural activities in the new lands. Construction of HAD basically confirmed the fact that the supply of arable land in Egypt is not necessarily inelastic, as was often assumed in the past. Nearly 650,000 acres out of a total of 805,000 acres of land reclaimed in the 1960s was made possible directly due to water from HAD.

The changes in Egypt's arable areas during the period 1897–1990, are shown in Table 6.2. It should be noted that between 1907 and 1980, the arable area increased by only about 700,000 acres, while the country's population increased nearly fourfold, from 11.2 to 42.1 million. The area of arable land available per person declined by 71 percent during this seventy-three-year period.

The most detailed analysis of land resources of Egypt was completed in 1986 under the Land Master Plan (LMP). This plan concluded that 2.82 million acres of land could be reclaimed by using the Nile waters. In addition, another 570,000 acres could be reclaimed by using the groundwater in Sinai and the New Valley. Thus the total land that could be reclaimed, subject to water availability, was estimated at 3.40 million acres. The LMP study considered land only for irrigated agriculture. Other uses of land like fisheries, forestry, and wildlife habitat were not considered. The LMP study divided the potentially reclaimable land into five categories depending on one or more land-use and management options. These options considered cropping patterns, irrigation and drainage systems, and farm types. More than half of the land proposed for reclamation is considered to be coarse to gravely sands.

The present estimate of cultivated area in Egypt is 7.49 million acres, of which 7.21 million acres are in the Nile Valley. Estimates of land-loss at present due to topsoil skimming and urban encroachment average 30,000 acres per year. It is essential that the government gives urgent attention to reduce the loss of arable land to urbanization for three important reasons. First, with increasing population, existing agricultural land areas should not be allowed to be lost. Second, land reclamation is an expensive process, hence it would be desirable not to lose any additional land that is already productive, and then try to compensate for that loss by reclamation. Third, often land lost due to urbanization is more productive than the reclaimed land. While the laws exist to prevent the loss of arable land due to urbanization, their enforcement is quite weak and erratic, as indeed is the case in many other countries.

Land reclamation in Egypt has been practiced over several thousand years. For most of this period, reclamation was concentrated primarily in the Nile Valley and the Delta, since land in these areas could be reclaimed with low levels of technology and investment. Impressive progress was made in land reclamation in the nineteenth century, at the beginning of which the cultivated area was estimated at 2 million acres, of which 250,000 acres could be cultivated only in the summer. By 1848, the area cultivated had increased to 2.6 million acres; by 1880 to 4.7 million acres; and by 1900 to 5 million acres. Thus, during the nineteenth century, the arable land area increased by 150 percent, or 3 million acres. Construction of the High Aswan Dam significantly increased both the supply and reliability of irrigation water. This, in turn, considerably hastened the process of land reclamation. Thus, between 1960 and 1971, a total of 912,000 acres of land were reclaimed, much of which was in the Western Delta. During the decade 1978–1988, an additional 74,000 acres were reclaimed.

There has been a major policy shift during the 1980s. The government became disillusioned with the overall performance of the state farms because of their inefficiency, inability to adopt new farming practices quickly, and the general lack of development of new farming systems more applicable to desertlike conditions. A policy decision was taken to allocate new lands in a varying ratio of 60:40—three-fifths to investors with adequate capital to develop their own farms; two-fifths to economically disadvantaged groups, unemployed graduates, and retired government personnel.

The total investment cost for land reclamation has been significant. For example, since the 1952 Revolution, over £E (Egyptian pound) 3 billion has been spent on land reclamation. The Land Master Plan study estimated that the investment cost for land reclamation varied from £E 3,000 to £E 7,000 per acre. In the remote areas, the high cost of infrastructure increased the cost to £E 8,000. The government has pledged not to dictate cropping patterns in these areas, and many farmers are planting high-value crops, probably perennial fruit crops, to get attractive returns on their investments. The potential impact of this policy on market prices of fruits needs to be investigated.

In Egypt, land can be productive only if water is available for irrigation. As population grows and achieves a better standard of living and more industrialization, water demands for the municipal and industrial sectors will increase. Since these two sectors are most likely to have higher priority than the agricultural sector, the future of reliable water supply for the reclaimed areas should receive serious attention. Even so, the share of water available to agriculture will decline steadily. Accordingly, efficiency of water-use in Egypt has to be increased to ensure that all the reclaimed lands will continue to receive their share.

7

Research Process, Human Resources, and the Quest for Sustainability

Howard A. Steppler

Sustainability is a relatively new concept in our lexicon. To couple it with research and human resource development is clearly to identify two essential components in the quest for sustainability. The concept of sustainability gained world-wide attention with the publication in 1987 of the UN report, *Our Common Future*. This report, underlining the fundamental need for sustainable development and the interdependence of nations in seeking that sustainability, opened with the sentence: "The earth is one but the world is not." Each nation's responsibility was unequivocally expressed: "Sustainable development can only be pursued if population size and growth are in harmony with the changing productive potential of the ecosystem."

First, a digression on my concern with sustainability and population growth. Food sufficiency can be attained by balancing a simple equation: food sufficiency exists when population x calories per person = food per land unit x total land units. I have used land units as the input factor. Clearly one could use the "limiting" resource (e.g., water) to reflect a more specific national situation, as in Egypt. In this equation two variables are essentially fixed, namely, calories required per person and total number of land units, where the latter can be considered on a national, regional, or world basis. The remaining variables, population and calories produced per unit of land, are variable and undergoing constant change. Our role as agriculturists is to maximize the right side of the equation. However, with no restraint on the left side (the variable population), it becomes increasingly difficult to balance the equation, let alone achieve sustainability.

With this digression, I return to consideration of the right side of the equation, in particular with respect to research and human resources. Three incontestable statements can be made about agriculture. First, the primary role of agriculture in society is to produce food for man and his animals. Second, all agricultural production activity is extractive: it inevitably consumes resources, whether they be in situ or imported. Third, the green, photosynthetically active plant is the key to the system. It drives the system.

The second statement leads me to the axiom that only a completely closed agricultural production system, in which all the elements are re-cycled—nothing removed—can in theory be sustainable. This is clearly an untenable proposition, so our challenge is to design systems which are sustainable yet tolerate removal of products for use outside the system.

There are many definitions of sustainability for agriculture. The report on sustainability by the Technical Advisory Committee (TAC) of the Consultative Group for International Agricultural Research (CGIAR) states that "sustainable agriculture should involve the success-ful management of resources for agriculture to satisfy changing human needs while maintaining or enhancing the quality of the environment and conserving natural resources" (TAC 1989). This is more a statement of characteristics and intent than a definition to provide a rigorously defined goal that can be tested experimentally. Lynam and Herdt (1988) discuss the problem of definition in some detail and propose a set of seven propositions to be considered when developing a research program to address sustainability. Huxley (1989) has compiled a study of sustainabil-ity in the context of agroforestry which contains a series of papers on the issues. It emerges from these papers, and from definitions used at the Alexandria conference, that a rigorous definition of sustainability is needed if it is to be a subject of scientific research.

The Research Agenda

Setting aside this question, I will address the question of determining the research agenda, and the concomitant requirement of human resources for the enhancement of the agricultural production system. The produc-tion inputs to the system which the agenda addresses can be classified.

1. Climatic components: chiefly water, temperature, daylength, and light intensity; light quality may well become a factor as a conse-quence of the reduction in the ozone layer
2. Edaphic components: both physical and chemical
3. Biological components: plants, animals, and their associate com-plex of pests, including soil organisms

The system functions in an environment influenced by social, political, and economic factors—the whole complex is managed by man. In an earlier study (Steppler 1978), I discussed these components with respect to their stability, transferability, and malleability or ease of manipulation. The analysis demonstrated that enhancement of a system's performance might come from changing the characteristics of components, changing the components per se, modifying the system when component change is ineffective, or from a combination of these actions. It also follows that change is more difficult as components become less malleable or transportable (more stable).

The research agenda can best be developed by considering the sequence of actions necessary to arrive at the agenda that can effect a change in the system, including the development and use of new technology. The following five stages in the process are suggested:

1. Develop a rigorous and unambiguous statement of the problems of the system, including a thorough analysis of all aspects. This is likely to establish a set of problems that demand the setting of priorities.
2. Identify the constraints, specify the nature of new technology or intervention and the point of entry into the system.
3. Seek new technology from among existing research findings and modify it if necessary for local use; or design and conduct the research necessary to develop new technology.
4. Determine the efficacy/appropriateness of the new technology by testing it under field conditions.
5. Arrange for and undertake the transfer of the proven technology to the ultimate user.

At the same time, there should be a clear determination of the support infrastructure to facilitate the optimal utilization of the technology and plans prepared to have the infrastructure set in place at the appropriate time. These five stages and the related infrastructure are shown in Figure 7.1.

The research agenda determined through stages 1 to 3 will obviously be concerned with one or more of the factors identified under the broad category of inputs. Further, it is important to recognize that some of the constraints are amenable to a research activity while others may require a different type of activity/instrument to effect change. This sequence of actions is, in my judgement, essential to the conduct of a research-cum-technology generation program irrespective of the area in which the research is undertaken. I would also suggest that most researchers use the sequence without realizing it, particularly when they are handling a commodity (for example wheat) with a disease problem. As one is confronted with more complex situations the need for a formal recognition and utilization of this approach becomes essential to developing appropriate technology. The demands for the sustainability of agricultural production systems clearly exhibit that complexity.

The attention paid to the first stage will set the tone for the whole research process. In the production system, in particular when sustainabil-

Figure 7.1 The Cycle of Development and Transfer of Technology

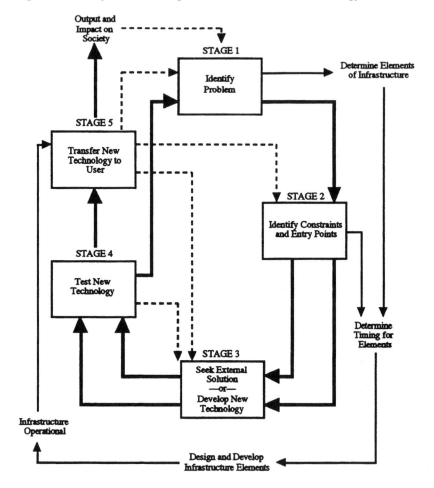

ity is an issue, one is faced with both internal and external factors. All of these should be considered when analyzing the system to develop the rigorous statement of the problem. At the very least, the biological and social sciences must be represented. The broad problem can then be disaggregated to a set of subproblems to which priorities may be assigned. The execution of the analysis requires a robust—even though imperfect—definition of sustainability. Without it, the second stage statement of the problem will become an illusion of rigor without the capacity for experimentation.

Stage 2 identifies the constraints and the points of entry into the system for the new technology. It should determine whether the constraints can be removed by research or whether they call for some other type of action, as for example, changes in government policy; changes in land tenure; provision of credit; or establishment of seed production units. Critical time paths for the provision of infrastructure should be considered at this stage.

There are two avenues for action in stage 3—either seek appropriate solutions elsewhere or develop your own solutions by research in situ. These are not mutually exclusive. The normal course is to seek the stop-gap, short-term solution from elsewhere while developing your own research program to generate the most appropriate technology. Whichever path is taken the most skillful scientists will be needed to design research with a high probability of success. We are dealing with a complex system, but it is most probable that research will be directed to components or factors recognizing the classification into the three primary components. Understanding the relationships between these components and production-management factors will obviously be a part of the research activity. In summary, the design process for the research agenda begins with the system, breaks up to clearly defined, researchable problems, then reassembles the system to examine the interactions between the modified components and determine the efficacy of the new system.

The ability to find appropriate technology or solutions outside your own program is predicated on three assumptions:

- That the researcher has been able to describe the desired characteristics of the technology; these should be in well-known terms—biological, physical, nutritional, chemical
- That the material offered be described in the same terms as those used by the researcher
- That there be a sense of where one has the highest probability of locating the desired material

Stage 4 completes the inner circle of Figure 7.1. It is the natural outcome of the research program—rigorous testing of the generated technology. Assuming that the first stage of the process was properly executed, the fourth stage should present few difficulties. The conditions which the technology was designed to ameliorate will have been identified, hence the appropriate test should be clear and the parameters of an acceptable solution specified. The validation of the technology must be done under the ultimate user's conditions. The necessary performance criteria for acceptance, and by inference rejection of the technology, should have been established in stage 2 prior to the initiation of the generation process. It should be clear from Figure 7.1 that we are dealing with a cycle and not with a straight-line process. The concept of continuous improvement in technology is inherent in the cycle.

Stage 5 involves withdrawing the proven technology from the generating system and transferring it to the final users. Figure 7.1 also shows that one would anticipate feedback both from the fifth stage and from the society. This feedback may impact virtually any stage of the cycle.

Success of the new technology is also related to the timely appearance of supporting infrastructure. This as suggested may be anything from social structure, to credit, production inputs, or market opportunities. All should be in place as the demand arises. Thus a critical time path should be prepared to ensure that these elements are operational as required. The nature of the infrastructure should first become evident in stage 1 and be further refined in stages 2 and 3.

It is possible to misinterpret Figure 7.1. The figure could be claimed to show that there is an inner cycle of research and technology generation that is self-contained and that can operate independently of the ultimate user: that the ultimate user (stage 5) receives spin-off from the research activity. I assure you, that is totally false, although I realize that this may have been perceived to have happened in some cases in the past. The figure shows the feedback from stage 5 and from society to the inner circle. The following discussion on human resources will show more clearly the relationships between the various stages and I hope a rejection of the potential trap.

Human Resource Development

There is no argument about the need for man's intervention in the process of agricultural production and in the generation of new technology. However, two important questions should be asked with respect to man's intervention in the generation and transfer of new technology:

- How do the two groups of professionally trained agriculturists, the extension agent and the research scientist, fit into the process and what is their role?
- What kind of education should each have to better fit them for their role?

Stage 3 is the domain of the scientist; stage 5 is for the extension agent. But what about states 1, 2, and 3? It is not so clear. Stage 1 is probably the most critical in the whole process. It is not the exclusive domain of any one group—a group that might then reflect its own discipline in problem identification. What is needed is a multidisciplinary group, acting as a team and counterbalancing each other's viewpoint. The extension agent has as much right to be here as the scientist; the same is true for the economist and the sociologist. All must direct their attention to the ultimate user, at the same time recognizing external pressures from society.

Identification and rigorous definition of the problem will define the domain to which the solution has the greatest relevance. Since the technology has been developed for a specific set of ultimate users, and since the transfer of the validated technology is the responsibility of the extension agent, it follows that the agent should be a member of the team in stage 4, along with all from the team of stage 1. Each has the responsibility to ensure that the revised system incorporating the new technology performs up to their expectations.

Stage 2 presents somewhat the same problem as stage 1. However, at this point the research agenda is being formed: it seems appropriate that the scientist would play the lead role, with the active participation of others. The timing for the development of infrastructure begins in this stage and people with related concerns should be part of the discussions.

Stage 4 can only legitimately validate those parts of the system that have been generated in stage 3. Other, external constraints may have been identified (in stages 1 and 2) but they will have resulted in changes reflected in the infrastructure. If this is the case, a second level of testing and validation will not take place until stage 5. Again, all disciplines should be involved in this penultimate approval, with feedback possibility. I deliberately use the term *penultimate*. The final approval comes with acceptance of the output of the system, and evaluation of its impact on the society—again with the provision for feedback.

It is easy to visualize two scenarios of differing complexities. One deals with a discrete problem of plant disease. The entire testing will be in stage 3 and final validation in stage 4. The other scenario, at the other end of the spectrum, is our problem of sustainability, treating the whole system internally and with a myriad of external factors. This would demand the full testing and validation of stage 5 and beyond. In either case, we are drawn to a process of continuous evaluation, feedback, and redesign of the system.

I drew attention to the need for a robust, quantifiable definition of sustainability. The cycle of development and transfer of technology is predicated on the assumption that such a definition is in hand. The definition would embrace at least three fundamental characteristics, namely, a quantifiable output; a time frame over which the sustainability performance is measured; and a measure of the impact on the natural resource base. The definition would be a holistic statement for the system, then a group of compatible subdefinitions nested within the overall definition. The holistic definition would be operative in stage 1; the subsets would begin to apply in stages 2 and 3; and the holistic would take precedence again in stages 4, 5, and beyond.

Research Education

Three potentially competing principles are applicable when debating the formation of our education curriculum to prepare people for a role in agricultural research. First, the individual must be prepared to undertake the role/job identified by the society. Second, the individual should be prepared to offer a particular skill, that is, to be trained in depth. Third, as the demands, problems, and priorities change, the individual should have a degree of intellectual flexibility—be able to adapt to these changes. The difficult problem is to arrive at the appropriate mix of the three principles when designing a curriculum. An overzealous addiction to any one of them can distort the program.

The first proposition is not generally welcomed by universities—they consider this to be too restrictive. Their approach is to prepare the individual for a research career in a discipline and even further a specialized area within the discipline. I believe that the urgency of many problems, coupled with the limited human resources that many countries can devote to a particular problem, calls for more specific preparation of individuals. This carries with it an obligation for the employer to provide time and resources to the researcher and the team to develop solutions.

The second proposition is the converse of the first. The university has tended to excel in this area, but with respect to discipline skills, not functional skills. I recognize that training in functional skills is necessary. I do not, however, consider it to be sufficient. In this chapter, I have repeatedly used the term *systems.* I strongly urge that one skill that should be acquired by all is a familiarity with the systems concept and some of the related research methodology. Each graduate student, particularly at the doctoral level, should be familiar with the concepts of analysis and interrelatedness of the various disciplines expressed in the Figure 7.1 cycle. This does not minimize the need for a sound preparation of the student in the discipline of choice. It does emphasize that the student be aware of its relationship to the overall system.

The third proposition is a plea for intellectual development of the individual—that there be a strong educational component in the curriculum. The student must be encouraged to develop an intellectual curiosity that goes beyond the chosen discipline. Students should be exposed to innovative thinking and imaginative approaches to problem solving— such as being given the challenge of working with a multidisciplinary group in a case study of the cycle of development.

I know of no models or guidelines that can be used to determine the best mix of these three principles. As a starting point, about one-third to one-half of the program could be devoted to the third principle. The remainder should be dedicated to the second principle—with the assumption that it responds to the challenge of the first.

References

Huxley, P.A. Editor. *Viewpoints and Issues on Agroforestry and Sustainability.* ICRAF. Working Paper, 1989.

Lynam, J.K., and R.W. Herdt. *Sense and Sustainability: Sustainability as an Objective in International Agricultural Research.* Paper for the CIP-Rockefeller Conference. Lima, Peru, 1988.

Steppler, H.A. "Natural Resources and Unsolved Environmental Problems," in *Distortions of Agricultural Incentives.* Edited by T.W. Schultz, Bloomington: Indiana University Press, 1978, 49–66.

Technical Advisory Committee (TAC). CGIAR. *Sustainable Agricultural Production.* Rome: FAO. Research and Technical Paper 4, 1989.

World Commission on Environment and Development. *Our Common Future.* London: Oxford University Press, 1987.

Part 4

Adaptation of Technology

8

Toward a Sustainable Rice Production System for Egypt

Mohamed Sayed Balal

Since Egypt's First Five-Year Development Plan was launched in 1982, efforts to increase food production have received top priority in the Agricultural Development Program (ADP). During the Second Five-Year Development Plan (1987–1992), the main objectives of the ADP were to increase food self-sufficiency; agricultural production to provide raw materials for industry and export; farm productivity through adoption of improved technology; and farmers' income and welfare.

During the 1980s, Egypt made significant progress in developing its national economy, and the agricultural sector played an important role. A major achievement of the agricultural sector was the 75 percent increase in the total output of cereals, from about 8 million metric tons in 1981 to about 14 million metric tons in 1991. The key factors behind this achievement were:

- *Political stability and commitment*: Strong commitment has been shown by the national leaders, from president and ministers down to village leaders.
- *Multidisciplinary research programs:* Three strong multidisciplinary programs have been established for major cereal crops—rice, corn, and wheat—to continue generating and improving production technologies. More emphasis was placed on varietal improvement, crop management, and integrated pest management.
- *Adaptation of technology:* An efficient system for adaptation of the improved technology was established in four phases: technology verification; technology dissemination; mass guidance by the National Campaign; and farmers' participation and feedback.

Within this context, this paper deals specifically with (1) recent developments in rice production in Egypt; (2) multidisciplinary rice research programs; (3) role of adaptation of technology in sustaining agricultural systems; and (4) the national strategy for improvement in rice production.

Recent Developments

In Egyptian agriculture, rice has economic importance for several reasons: It is a preferred food of most Egyptians, contributing about 20 percent to per capita cereal consumption. Rice is also an important export crop:

1970	*1980*	*1987*	*1991*
400,000	25,000	70,000	200,000(tons)

Rice occupies about 420,000 hectares, or about 20 percent of the cultivated area, during the summer season; it also consumes about 10 billion m³ or about 18 percent of total water resources; rice farming engages about one million families, or 10 percent of the country's population; and its average annual value during the period 1985–1987 amounted to about LE 1 billion—about 12 percent of the average annual value of crop production and about 8 percent of the total value of agricultural production.

Because of the fertile soils of the Nile Delta, high intensity of sunlight, few diseases and insect pests, and warm weather and a good irrigation system, rice yields in Egypt have been among the world's highest: the average yield of 7 tons/ha is more than double the world average of 3 tons/ha. The average yield of rice in Egypt increased from 5.7 tons/ha during 1984–1986 to 7.1 tons/ha in the 1989–1991 period. It was 7.3 tons/ha in 1990 and 7.5 tons/ha in 1991, which may be the world's highest (Table 8.1). Two of the seven rice-growing governorates (Beheira and Gharbia) yielded about 8 tons/ha. These high yields were achieved by:

- Releasing and spreading new high-yielding varieties: Giza 175; Giza 176; Giza 181; and IR 28
- Transferring appropriate technology to the farming community to improve crop management
- Monitoring production constraints and farmers' problems, with prompt follow-up action under the umbrella of the National Rice Campaign

Studies conducted to determine the yield potential or the "yield gap" showed that yields of the demonstration fields—using the best-recommended technology and improved high-yielding varieties—averaged 10.3 tons/ha during 1989–1991. Yields of the demonstration fields exceeded the national average yield by about 45 percent. Results in the 1991 season showed that the potential yield ranged between 7.7–13.0 tons/ha. The gap

Table 8.1   Rice Area, Production, and Yield in Egypt, 1984–1991

Year	Area		Production		Yield	
	Hectares 1,000	Index	Tons 1,000	Index	Tons/ha	Index
Average 1984–86	420	100	2,400	100	5.71	100
1987	414	98.57	2,413	100	5.83	102
1988	360	85.71	2,182	91	6.06	106
1989	413	98.33	2,668	111	6.47	113
1990	435	103.57	3,167	132	7.28	127
1991	454	108.10	3,411	142	7.51	132

between the potential yield in the demonstration fields and the national average can be attributed to soil salinity and alkalinity; inappropriate water management; the large area (about 60 percent) planted to traditional japonica varieties; spread of blast disease; inappropriate pest management; and a high percentage of postharvest losses.

Multidisciplinary Rice Research Program

With limited land left for horizontal expansion and scarce water resources, we must increase yield and productivity of rice through a well-organized interdisciplinary team approach. Although Egypt is one of the few countries with average yield of more than 7 tons/ha, it is possible to achieve 12 tons/ha, as yielded on experimental plots and in demonstration fields. To encourage more farmers to adopt the improved production techniques, an interdisciplinary research program was established in the early 1980s. The program includes:

- Plant breeding—development of new improved varieties with resistance to diseases and insect pests, early maturity, and short stature
- Seed production—putting pure seed of the new varieties into farmers' hands
- Agronomy, including plant nutrition, water management, and cultural practices—maximizing yields
- Plant protection against weeds, diseases, insects, and other pests
- Mechanization—developing small-scale implements that can be locally manufactured and maintained

- Economics—keeping in mind that successful new technologies usually cut costs
- Extension services—verification and transfer of new technologies to farmers

To strengthen the rice research program, the government has established the Rice Research and Training Center (RRTC) at Sakha, Kafr El-Sheikh Governorate, with strong support from the Ministry of Agriculture and Land Reclamation (MALR); the International Rice Research Institute (IRRI), with which we have been collaborating for about 20 years; and USAID, which has been providing funds for the Rice Research and Training Project through the National Agricultural Research Project (NARP)-IRRI contract.

The RRTC was established in January 1987. The center has a full range of well-equipped research facilities. In addition to Sakha, it has three testing stations at Gemmiza, Zarzoura, and Sirw, and twenty on-farm verification sites in the seven rice-growing governorates, namely: Kafr El-Sheikh, Dakahlia, Beheira, Sharkia, Gharbia, Damietta, and Fayoum. The Rice Research Program also employs about ninety rice production advisors to help disseminate the improved technologies from Sakha to the other districts.

Recent yield increases are a result of the following achievements:

- Release and spread of the short-stature, early-maturing, high-yielding varieties: IR 28; Giza 181; Giza 175; and GZ 2175 (Giza 176)
- Better nutrient management through efficient use of the nitrogenous fertilizers and application of zinc sulfate to rice nurseries
- Spread of chemical weed control from about 12 percent of the rice area in 1981 to about 60 percent in 1990
- Increase of the seed renewal rate from 50 percent in 1981 to about 70 percent in 1990
- Use of integrated pest management, including genetic resistance, cultural practices, and chemical control
- Strengthened relationship among research, extension, and rice growers—the efficient National Rice Production Campaign

Recently four improved varieties have been released for general cultivation: IR 28 was introduced from IRRI. It was released in 1985 because of its early maturity, short stature, and blast resistance. Giza 181 was released in 1986 due to its high yield, blast resistance, and excellent long grain. Giza 175 combines the improved indica plant type (IRRI) and japonica grain shape with early maturity and blast resistance. Giza 176, a typical japonica variety, was released in 1989. Because of its high yield, resistance to blast, and acceptable grain quality, it accounted for 30 percent of the rice area in 1991.

In the 1990 and 1991 seasons, the relative productivity of the improved varieties was tested and the results show that the new varieties have 20–25 percent higher yield-levels than the traditional japonica varieties. Presently, these improved varieties are grown on about 40 percent of the rice area in Egypt. It is expected that the area planted to the improved varieties will expand and cover more than 80 percent of the rice area by 1994.

The Role of Adaptation of Technology

Agricultural sustainability has been defined and described in many ways. But all point to one dynamic concept: the growing needs for agricultural production should be met without degrading the natural resource base on which agriculture depends (DeDatta 1990). Harwood (1990) has defined sustainable agriculture as one that can evolve indefinitely toward greater human utility, greater efficiency of resource use, and a balance with the environment that is favorable to mankind and other species. He has further summarized several common themes for development as follows:

- Food production must continue to increase to meet the demand of rapidly expanding populations (about 2 percent per year)
- Total agricultural employment as well as individual income from agriculture must expand greatly
- Efficiency of use of capital, land, and production inputs must increase sustainability
- Production systems must be structured for the lowest possible use of pesticides

According to Manwan (1990), and as shown in Figure 8.1, a sustainable production system depends on four major factors with very close relationships: government policy; improved technology; external support; and farmer's participation.

As mentioned earlier, political stability and the commitment of the country's leaders—establishing facilities and allocating budgets—has played an important role in sustaining agricultural production. It is important to have a national strategy with clear production goals and objectives for each commodity.

Research output could be in the form of information and technology related to the production, economic, and social aspects. Production technology should be improved with more emphasis on breeding high-yielding, early-maturing varieties with resistance to the major diseases and insect pests; on crop management to maximize yield of the improved

Figure 8.1 Factors Affecting a Sustainable Farming System

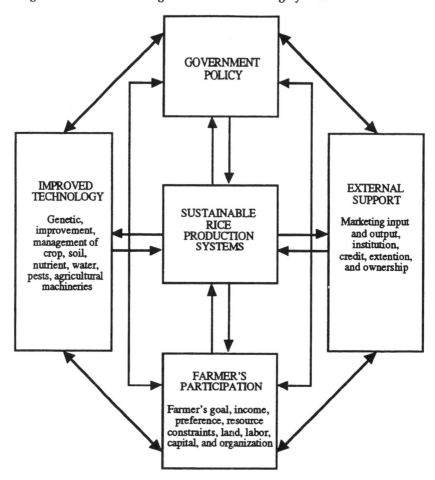

varieties and increased efficiency of irrigation water and fertilizers; and on integrated pest management to control weeds, diseases, and insects.

Marketing inputs and outputs are important factors affecting sustainable systems. The major inputs such as seed and fertilizers should be available to farmers at the appropriate time. Also, the floor and ceiling prices should be determined and announced at an early time before harvesting. Farmers' participation and acceptance of the new technology is the aim of the technology adaptation program. It is the most important factor affecting sustainable systems. In our Rice Research and Development Program the technology adaptation component (Figure 8.2) is carried out in four phases.

Figure 8.2 Scheme of the Rice Technology Adaptation Program

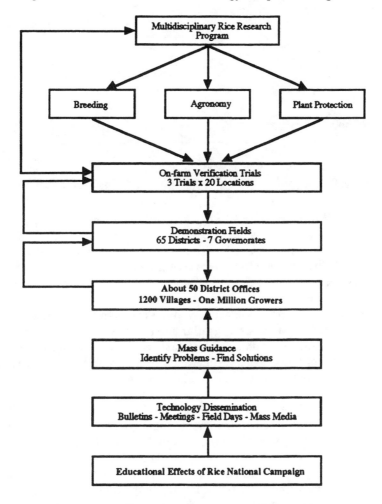

1. *Technology verification:* Three on-farm verification trials at twenty locations are conducted every year to confirm the new research findings in farmers' fields.
2. *Technology dissemination*: About sixty demonstration fields of about two hectares each are established, where the package of recommended practices is applied using the new varieties. In 1991, yields of forty-eight demonstration fields ranged between 7.7 and 13.0 tons/ha with an average of 10.3 tons/ha, which was about 45 percent higher than the national average.
3. *Mass guidance (National Rice Production Campaign):* Recently

Table 8.2 Rice Production Strategy, 1992–1997

Period	Area		Yield		Production	
	Thousand feddans	percent	Tons per feddan	percent	Million tons	percent
1984–1986 (Base Period)	1,000	100	2.40	100	2.40	100
1989–1991	1,033	103	3.00	125	3.10	130
1992–1997	1,200	120	3.50	146	4.00	167

the Ministry of Agriculture, in coordination with the Egyptian Academy of Science and Technology, organized a number of national campaigns for major crops such as rice, corn, and wheat. The scheme is designed to provide conditions where a large number of farmers adopt the new technology to increase productivity and income. Basically the system is focused on motivating farm communities to participate actively in the program for increased production. Beside mass guidance, the system also monitors production constraints and farmers' problems in the season, from land preparation to harvesting, and various agencies act to alleviate these problems. The campaign is active in all rice-growing governorates.

4. *Feedback:* Every year, the Agricultural Extension and Rural Developmental Research Institute of the Agricultural Research Center (ARC) investigates the educational effects of the National Rice Campaign to determine the degree of farmers' participation and to recognize constraints. For the 1991 season, the study showed a high degree of farmers' participation for both demonstrators (89 percent) and other farmers (70 percent). Giving feedback, the study showed that farmers faced nineteen constraints, such as lack of irrigation water, high prices of fertilizers, nonavailability of appropriate pesticides, and seed quality.

National Strategy for Improved Production

During the Second Five-Year Development Plan, rice production increased from 2.4 million tons during the base period (1984–1986) to 3.41 million tons in 1991, or by 42 percent. This large increase in rice production reflected a 32 percent increase in productivity (7.51 tons/ha) achieved in

1991 (Table 8.1). To keep pace with nearly 2.7 percent annual population growth, while exporting about 250,000 tons annually, rice production has to increase from the present 3.1 million tons to about 4 million tons by the year 2000—an increase of 30 percent. To achieve this goal, a multipronged strategy has been developed as part of the Third Five-Year Development Plan (1992–1997). The strategy, aimed at increasing productivity per unit of land, water, and labor, and increasing farmers' income, will:

1. Support the multidisciplinary Rice Research Program to continue generating improved technology. Emphasis will be on four areas: accelerating the varietal improvement program to develop new japonica varieties with durable resistance to blast and early maturity; intensifying research on crop management to maximize productivity of the improved varieties and increase fertilizer and water-use efficiencies; intensifying research on integrated pest management to control weeds, diseases, and insects with minimum use of pesticides; and strengthening the collaboration with the International Rice Research Institute (IRRI) and other international organizations.
2. Support the technology adaptation program in all phases: from technology verification and demonstration to mass-guidance motivation of the rice-farming community.
3. Improve productivity by expanding the area planted under improved high-yielding and short-duration varieties and by adoption of new technologies.
4. Identify the less productive and problem-soils districts. About 20 percent of rice soils are salt-affected and their productivity is about 40 percent less than normal soils.
5. Assure the availability of irrigation water to expand the area under rice in the Nile Delta by 20 percent by the year 2000.

References

Balal, M.S. *Rice Varietal Improvement in Egypt.* Proceedings of the International Conference, "Rice Farming Systems." Sakha, Egypt, January 1987.

———. "Rice Production Status and Strategies in Egypt," presented at the Seventh National Rice Research Conference. Sakha, Egypt, March 1990.

Brady, N.C. "Making Agriculture a Sustainable Industry." Proceedings of the International Conference, "Sustainable Agricultural Systems," Soil and Water Conservation Society. Iowa, 1990: 20–32.

DeDatta, S.K. "Sustainable Rice Production: Challenges and Opportunities." 17th Session of the International Rice Commission, FAO, Goiania, Brazil: February 1990.

Harwood, R.R. "A History of Sustainable Agriculture." Proceedings of the International Conference, "Sustainable Agricultural Systems," Soil and Water

Conservation Society. Iowa, 1990: 3–19.

Manwan, Ibrahim. "Increasing Rice Production in Indonesia: Policy, Strategy and Achievement," Network for Genetic Evaluation of Rice. Bogor, Indonesia: December 1990: 10–15.

9

Transfer and Adoption of Horticultural Technology: A Global Overview

Hamdy M. Eisa

Horticultural crops—usually referred to as high-value crops—are assuming an increasingly important role in the economy of many developing countries. This should be tempered with the fact that most of these crops are highly perishable. They are of high value only if produced in a timely way and marketed and sold at the right price. Horticultural crops are an important source of cash to farmers and contribute significantly to increasing their incomes. The integration of the production, processing, distribution, and marketing chain is crucial. The transfer and adoption of suitable horticultural technology, at each stage from production to marketing, is of paramount importance to assuring a reliable supply of high-quality horticultural products at prices that are both profitable to producers and reasonable to consumers.

Horticulture and Health

Recently, the contributions of fruits and vegetables to improved human health have been recognized. They supply anticarcinogenic fibers (soluble fibers), vitamins, and minerals, and help to reduce amounts of natural carcinogens in people's diets. The role of fruits and vegetables in lowering the risks of cancer and heart diseases is being accorded higher visibility by the medical communities and health and consumer advocacy groups. This trend is more pronounced in developed countries, but consumers in developing countries, too, are becoming more aware of the dietary importance of fruits and vegetables. More effort is needed to make everyone aware of the beneficial contributions of horticultural crops. Information should be given to both urban and rural populations; to both affluent and poor (Eisa and Quebedaux 1990).

Herbs and spices have always played an important role in the palatability and acceptance of food. Increased communication, travel, and exposure to different cultures have contributed to the promotion of ethnic cuisines and herbs and spices. Promotion of natural additives and their preference over synthetic chemicals have contributed to increased demand. However, the elasticity of demand for herbs and spices is lower than that for fruits and vegetables: herbs and spices are consumed in relatively small amounts. Medicinal plants have always played an important role in human health and recently the use of herbal medicines has increased in many developed countries. This is due to health awareness, the promotion of organically grown food, and increased affluence, particularly of urban consumers. Floricultural and ornamental crops have assumed greater importance, improving the aesthetic environment, particularly in congested urban dwellings. Appreciation for gardening in alleviating urban stress, as therapy for the handicapped, and helping in the speedy recovery of patients has also been scientifically documented.

This paper will examine the role of horticultural crops (primarily fruits and vegetables) in the world economy, and new horticultural technologies for production, processing, distribution, and marketing. It will provide examples for adoption of these technologies, particularly in developing countries, and will highlight interlinkages in the horticultural development process between producers and consumers in domestic and export markets, with special reference to Egypt's drive to capitalize on its comparative advantage in horticultural crops.

Global Role of Horticulture

Most of the studies on the challenge of meeting increased food demand due to population increase have concentrated on cereals, leguminous crops, and livestock products. Little attention has been paid to horticultural crops. This may be due to the perception that the latter make a relatively small contribution to energy and the daily caloric demand of human beings. Since diversifying the diet serves the ultimate purpose of meeting the basic health standards, horticultural crops should assume a larger contribution in providing the basic (minimum) health criteria established by the international and national health organizations, planners, and policymakers. The growth in demand of horticultural crops would at least be comparable to the average population growth rates world-wide. Additional factors of increased demand with high income and awareness of contributions to human health would lead to increased consumption. Therefore, developing countries should not only base their horticultural production and marketing strategies on economic grounds and generation of foreign exchange but also on improving the health and nutrition standards of their citizens. Governments' strategies of poverty alleviation and meeting the basic food needs should include balanced nutrition and health improvement by providing adequate and timely supply of fruits and vegetables.

Horticultural Crops in the World Economy

In recent years, policymakers, analysts, and development specialists have expressed broad interest in the potential contribution of horticultural products to agricultural diversification, employment generation, and foreign exchange earnings in developing countries. However, little research has been done on world trade in horticultural products. This may be partly due to the multiplicity of primary, semiprocessed, and processed forms of horticultural crops and the difficulty in aggregating the data on a commodity basis (e.g., cereals, livestock, or oil products).

The developed countries provide the largest import market. From 1983 to 1985 they accounted for 85 percent of world imports of horticultural products. Similar trends were reported for the period 1985–1990. The developing countries' share of world imports in 1983–1985 (about 17 percent) is expected to increase to 20–23 percent in the year 2000. Therefore, future prospects of horticultural exports of developing countries will depend primarily on the growth of import demand mostly in the developed countries. The entrance of Spain and Portugal (large horticultural producers) to the European Economic Community (EEC) has contributed to increased European self-sufficiency of fruits and vegetables, i.e., exceeding 80 percent. The formation of a united Europe will undoubtedly affect trade with other countries.

Given the high per capita consumption of fruits and vegetables today and the projected slow rates of growth of income and population in developed countries, the annual rate of growth in their aggregate domestic demand in the 1990s probably will not exceed 1.31 percent for fruits and 1.08 for vegetables. The developing countries' exports of horticultural products are estimated to increase at rates ranging from 1.6 to 3.4 percent a year—in absolute value to US$12–15 billion in 1983–1985 prices by the year 2000 (Islam 1990). However, these aggregate forecasts should be taken with extreme caution due to the introduction of open market economic policies, and political changes, in the former USSR, in Eastern European countries, and in a unified Germany. This will undoubtedly change international trade patterns, including those for horticultural crops, and new export opportunities to developed countries may emerge.

Countries differ in their comparative advantages; likewise in their trade performance. In addition to basic agroecological endowments, there are differences in labor, land, and water productivities—resulting from research and development efforts—and these play a significant role in successful horticultural development strategies. The organization of an effectively integrated system of production, processing, packaging, transportation, storage, distribution, and marketing, both nationally and internationally, is crucial to success in horticultural exports. Economies of scale promote these activities by reducing their costs significantly, and the price advantage helps exports. Therefore, successful use of innovations (technological change) should ultimately lead to a reduced cost of the marketable unit (Eisa 1986).

Horticulture and Sustainability

The National Research Council of the United States defines sustainable agriculture as

> the integrated system of plant and animal production practices having a site specific application that will, over the long term (1) satisfy human food needs; (2) enhance environmental quality and the natural resource base upon which the agriculture economy depends; (3) make the most efficient use of nonrenewable resources and on farm resources and integrate, where appropriate, natural biological cycles and controls; (4) sustain the economic viability of farm operations; and (5) enhance the quality of life for farmers and society as a whole (Adams 1991).

Horticultural development is not different from other agriculture endeavors with regard to sustainability and balanced use of natural resources. The integrated life and resource cycles, whether biological or not, should be recognized. Their roles should be assessed in the development process. In this respect, the role of natural and human resources in sustainability of horticultural development, technology transfer and adoption, and the economic welfare of beneficiaries is of paramount importance. Natural resources to be assessed include land suitability (particularly arable land), rainfall amounts and distribution, periods of drought, rivers, soil conservation measures at the farm and country levels, and climate (temperature, relative humidity, sunshine, wind velocity, etc.). In addition, the contribution of energy resources is gaining increased recognition, particularly when modern inputs such as agrochemicals (fertilizers, pesticides, herbicides, and growth regulators), plastics, and farm machinery are to be used.

As for human resources, availability of labor (skilled and unskilled) and adequately trained manpower is essential in all phases of the production-to-marketing chain. Education, training, and technical assistance are closely linked to human resource development. The level of farmers' education is extremely important in accelerating the adoption of knowledge-intensive horticultural development programs. The role of women in development in general, and horticulture in particular, should receive serious consideration in production, processing, and marketing.

Innovative Technologies

The integrated contribution of art and science to the horticultural industry dates back to even before the hanging gardens of Babylon. This inseparable integration is characteristic of all facets of horticultural development and will remain so. Art in horticulture involves vision, taste, appreciation, innovation, sensitivity—a multitude of perceptions and conceptualization to develop plants and utilize them for the welfare of human beings. I will not, here, deal with the artistic side of the horticultural picture, though it is a major and a basic determinant in horticultural development, but will address recent horticultural technology innovations, transfer, and adoption.

Production Advances

Technology innovation through genetic means has progressively been explored and articulated since Mendel's genetic studies on garden peas. Seeds and vegetative propagules are considered the basic blocks for transferring innate genetic characters (genic or cytoplasmic) from one generation to the next. With the advent of hybridization, horticultural breeders have capitalized on heterosis, uniformity, disease and insect resistance, quality improvement, high yields, and so forth. They have developed unique horticultural crops through sexual and asexual means. The incorporation of beneficial horticultural characteristics (e.g., disease and insect resistance, long shelf-life, parthenocarpy) has opened the door to the development of unique varieties, such as greenhouse cucumbers (European cucumbers). Hybrid cucumber and tomato seeds are not sold by weight but by count. In spite of the high cost of hybrid seeds, farmers are willing to pay for hybrid seeds because the financial benefits and economic returns are higher than open-pollinated varieties. This does not mean that every hybrid development program will be successful and result in financial rewards.

Notable achievements in horticultural production technologies have been transferred and adopted in several developed and developing countries. Most of these innovations originated in developed countries, but their adoption in developing countries has included the making of some modifications to maximize benefits and make technologies relevant, e.g., the use of plastics in greenhouses, mulching, and row covers. The cohesive integration of these technologies is of paramount importance and an examination of the separate phases of technology innovations is needed to formulate sound recommendations. However, the ultimate criteria for the introduction of technologies are the financial and economic returns to farmers and the benefits to consumers.

Recent advances in seed production, conditioning, packing, and dis-tribution have led to wider distribution of improved seeds of vegetables and flowers. The combination for seed priming, treatment, coating, pal-leting, and, recently, pregermination and fluid drilling have contributed to the achievement of maximum uniform quality of the final product. Special seed drilling equipment and land preparation practices have been developed to place the high-quality seeds at the proper soil depth. Mini-mizing the internal seed moisture content to a level that is safe for each type of seed, then packing it in hermetically sealed moisture-proof cans, packs, or pails, is becoming a normal practice. Seed labels on many vegetable seeds provide information on germination (which reaches 99 percent), pest resistance, and horticultural characteristics. Yields of 300–400 tons/ha for greenhouse (European) cucumbers are not uncommon in the Netherlands. Similarly, high yields of greenhouse hybrid tomato have been obtained under controlled environmental conditions.

The introduction and adoption of tissue culture is becoming a stan-dard technique for fruits, vegetables, ornamental plants, and some tree crops (oil palm) in both developed and developing countries. The produc-tion of virus-free seedlings is important in producing healthy plants, particularly crops such as strawberries, potatoes, bananas, and orchids. Shoot-tip grafting in fruit crops (e.g., citrus) is an improved technique which has spread to India, China, Chile, Brazil, and Egypt. However, its use has to be linked with an effective program in budwood mother-tree maintenance, virus indexing, and budwood registration. Advances in rootstock evaluation are leading to the development of different degrees of dwarf trees in apples. Recent advances in biotechnology, including genetic engineering and protoplast fusion, will undoubtedly make incor-poration of the desirable characters in a breeding program more feasible and faster than present conventional means (backcrossing).

Irrigation and Fertilization

The improvements in combining irrigation and chemical fertilization, i.e., fertigation, have led to a revolution in effective saving of water resources and timely delivery of essential elements to the plants. Ex-amples abound of efficient fertigation methods and their effect on yield-levels. The metering of nutrient solutions in hydroponics, sand culture, nutri-film culture, and other media is being used extensively in many developing countries. Subsurface drip-irrigation (a relatively new method in which drip-irrigation laterals are buried permanently 20–60 cm below the soil surface) is being advocated in production of processing tomatoes. It is claimed that commercial yields of 200 tons/ha of processing tomatoes could be achieved in the San Joaquin Valley of

California, using a subsurface drip-irrigation system that accurately delivers water and nutrients (Phene et al. 1992).

Recent developments in fertilizer formulation, application, and marketing are numerous. The promotion of open market economies in agriculture, removal of pesticide and fertilizer subsidies, and further private sector involvement will undoubtedly change fertilizer production and marketing and eventually will have a positive impact on their role in horticultural development. Another growing part of horticultural development is the promotion of organic farming and specialized consumer demand for organically grown produce, particularly fruits and vegetables. Marketing opportunities for organic produce have expanded in developed countries and producers should explore these technologies.

Production Practices

Innovations in pruning techniques, cultivation, and harvesting are also fast developing. The use of machines in developed countries has been promoted to compensate for labor shortages and the high cost of skilled labor. Large-scale horticultural farming operations in developing countries, e.g., in Mexico and Chile, are adapting farm machinery to their needs. Harvesting, field handling, and reduction of field-heat are important operations that should be carefully integrated to minimize damage to produce. Use of vacuum cooling, particularly for leafy vegetables, is receiving increased attention.

Crop Protection

The judicious use of agrochemicals, particularly fertilizers, pesticides, herbicides, and growth regulators, is receiving increased attention from consumers and environmental groups. The promotion and application of the Integrated Pest Management (IPM) approach in horticultural production and marketing systems is being extended from developed to developing countries. These knowledge-intensive technologies require some degree of farmer literacy and training in the identification of economic thresholds for pest damage. The use of special equipment (particularly orchard sprayers) and its maintenance needs to be considered prior to commercial promotion. The suitability of this equipment to the scale of operations in developing countries should be assessed, under local conditions, before it is used commercially.

Marketing and Shipping Technologies

Recent advances in technologies for postharvest handling include the promotion of environments with controlled atmospheric conditions.

These can be created through gaseous exchange and increased carbon dioxide and nitrogen content. This technique is being extended to retail packages. The use of chemical preservers, additives, and irradiation is increasingly discouraged. Freezing and dehydration technologies are being used in fresh vegetables and dehydrated fruit mix at retail stores. Special varieties for snack foods are being developed. Transportation technologies and methods are very important in horticultural marketing. The cost of transportation, particularly by air, is crucial in making export marketing decisions. Marketing innovations, including promotion and presentation of produce at retail level, are important—and not only to the presentation of the product. They also affect the producer-country's image.

Chile, China, and Egypt

Several developing countries have carved out special market niches for their horticultural exports and established a sound reputation in specific commodities. This chapter cannot cover all these efforts, but I will briefly report on three countries: an established industry (the Chilean experience); an emerging industry (the Chinese experience); and a look into the future for Egypt.

The Chilean Experience

Commercial fruit production in Chile (Gnaegy 1992) began in the 1930s and blossomed in the 1950s with the advent of technological advances such as refrigerated transportation and customized methods of irrigation. Government policies at that time complemented to production and marketing for export. But in the 1960s and early 1970s, government policies turned against agriculture. With increased implicit taxation and the nationalization of industries, and with restrictions on exports—combined with blockades imposed by other nations as a way to apply international political pressure—the agricultural sector stagnated. Internally, political instability, economic uncertainty, and threats to private property combined, delaying investment in this potentially lucrative industry (Wylie 1990).

With the overthrow of the Allende government, many of these stifling factors were eliminated. The capital markets were freed up and allowed to find international parties, making them competitive; the regulations governing investments were in turn eliminated or simplified. Land rights became more secure. As key services such as communications and transportation were privatized, their costs became comparable with international costs, enabling producers to weigh the potential profit of their goods on the world market. These measures, combined with a highly-educated industry and export work force, allowed Chile to take the lead in the production of high-value fruits and vegetables for exports (Wylie 1990).

Chilean producers and farmers (I use the term loosely: most are investors who have various business interests, among which farming is merely one) have demonstrated both willingness and strong desire to search out, experiment with, transfer, and adapt advanced horticultural technologies from California to the agroclimatic conditions of Chile. Not only have these farmers experimented with varieties, horticultural practices, cooling and storage techniques, and pest and disease management, they have shared their knowledge freely among themselves in cooperation with the National Research Institute, which contributes to research as well as to the extension of this technology (Wylie 1990).

It is quite clear that in Chile the absence of government intervention was itself the primary reason for the dramatic success of this industry. The few government programs or regulations that do exist have been established to encourage the industry. For example, in 1985 the government created an export "drawback" program, whereby rebates are offered to exports of high-value fruits and vegetables whose free-on-board value equals less than US$2.5 million a year. As the value increases, the rebates are scaled back. They are eliminated once the value of that particular commodity equals US$18 million or more (FAS 1992).

For its part, private industry takes primary responsibility for its needs. The producers maintain a voluntary fruit export quality program, ensuring compliance with importing countries' regulations. Their largest importers are the United States, Japan, and the EEC, all of which have relatively high phytosanitary standards. They put a lot of effort into influencing the Ministry of Foreign Affairs to establish and maintain free trade agreements. Incredibly, the industry even funded a highway, open to the public, running from the main fruit and vegetable producing region of the country to the main port of embarkation (Wylie 1990).

China—Emerging

In China, the fruit production area, which occupied 22 percent of the total land used for industrial crops (orchards, cotton, hemp, sugarcane, tobacco, silk, tea, and oil-bearing crops), increased from 1.7 million ha in 1978 to 5.1 million ha in 1988. Production increased rapidly from 6.6 million tons in 1978 to 16.7 million tons in 1988. Yields of mature orchards also grew significantly: a rough estimate is that they grew at about 4.4 percent per annum during this period.

This rapid expansion of fruit production is largely due to the major economic reforms which improved producer incentives through the production responsibility system, decontrol of fruit marketing and prices, and encouragement of regional specialization (Eisa and Jaisaard 1990). Previously, priority had been given to production of staple foods—to the virtual exclusion of other crops. The new policies have encouraged the development of cropping matched to local agroecological conditions. These policies have been aimed at improving the incomes and living standards of small farmers and providing the consumers with a more varied and higher quality diet. During 1978–1988, market fruit prices rose at an annual average rate of 10 percent, considerably faster than the rate of inflation, indicating that the growing demand was not being met at prevailing prices. High and rising prices are major incentives for planting new orchards.

Several new varieties of fruits and vegetables have been introduced in China. However, resolution of patents and plant breeding rights is delaying the introduction of joint venture seed production companies. Biotechnology, including tissue culture techniques and genetic engineering, are being combined with shoot-tip grafting to produce virus-free seedlings of many crops and citrus budlings. The introduction of modern postharvest technology is slow: most of the consumption is still in fresh form and consumers' purchasing power increases only slowly. An exception is in the coastal and southern provinces near Hong Kong, where modern processed foods are available.

Horticulture in Egypt

Egypt's agricultural and nonagricultural exports declined in the 1980s. During that decade, agriculture's share in total exports also declined, from 22.5 percent in 1981 to 20.3 percent in 1989, with a peak of 24 percent in 1984. Cotton, oranges, and potatoes are the major agricultural exports. These three commodities constituted about 90 percent of the value of total commodity exports in 1989/90. Cotton, in spite of a steady decline in its export, still constitutes about 60 percent of total agricultural commodity exports, or US$222.25 million in 1989/90. In recent years, only 15 percent of production of these commodities has been exported, indicative of increased domestic demand (due to increased population and income growth). With severe limitations on arable land and water resources, and the rapid growth of population, increased production of horticultural crops, primarily for export, compared with the production of cereals and oil crops, will have to be decided by relative international prices, and whatever value-added benefits there might be for horticultural exports.

Since 1986, the government has adopted a number of important trade and foreign exchange liberalization policies affecting the agricultural sector. These policies have significantly helped agricultural exports. Their impact on import substitution of grains has also been substantial. Exports of dates, tomatoes, lemons, and limes more than quadrupled in the 1980s. Exports of onions and potatoes also increased at fairly high rates: the share of onions in total agricultural exports increased from 5 percent in 1980/81 to 9 percent in 1989/90; for potatoes, from 10 to 15 percent.

Horticultural production in the new lands has contributed to increased availability of grapes and peaches. The newly reclaimed areas offer great potential for horticultural production in general, but any such development should be based on an accurate assessment of the demand in domestic and foreign markets. Newly introduced and improved varieties and irrigation technologies (drip and sprinkler) should be incorporated into an integrated production-to-marketing system. Presently, the emphasis is mostly on separate components of production and processing, without accurate assessment of the total picture.

The emerging new markets in Eastern Europe and the former USSR need a different market approach, long-term strategy, and early penetration. Exports should grow with the development of open market economies in these countries. Egypt's comparative advantage vis-à-vis other Mediterranean suppliers—Israel, Turkey, Cyprus, Morocco—to the EEC should also be assessed. Egypt should intensify efforts to obtain favored country status with the EEC.

References

Adams, J.A. "Sustainable Agriculture: Applications for Citrus," *Citrus Industry.* 1991: 72(10) 30–33.

Eisa, H.M. "A Horticulturist's View of International Agriculture Strategies," *Hortscience,* 21(3): 378–381.

Eisa, H.M., and R. Jaisaard. "Horticultural Crops in China." Unpublished report. Washington, D.C.: The World Bank, 1990.

Eisa, H.M., and B. Quebedaux (editors). "Horticulture and Human Health: Contributions of Fruits and Vegetables Symposium," *Hortscience.* 1990: 25(12) 1474.

Foreign Agricultural Service (FAS), USDA. *Deciduous Annual Report.* Report No. C1 2002. Santiago, Chile: USA Embassy, January 1992.

Gnaegy, S. "Innovation and International Competition in Agriculture: The Case of High-Value Exports." Unpublished paper. Washington, D.C.: The World Bank, 1992.

Islam, N. *Horticultural Exports of Developing Countries: Past Performance, Future Prospects, and Policy Issues.* Research Report No. 80. Washington, D.C.: International Food Policy Research Institute, 1990.

Phene, C.J., R.B. Hutmacher, and K.R. Davis. "Two Hundred Tons Per Hectare of Processing Tomatoes—Can We Reach It?" *Horticulture Technology.* 1992: 1(1) 16–22.

Wylie, A. "Agricultural Development and Technology: The Growth of Chile's Fruit and Vegetable Export Industry," in *Sharing Innovation: Global Perspectives on Food, Agriculture and Rural Development.* Edited by N.G. Kotler. Washington, D.C.: Smithsonian Institution Press, 1990.

Part 5

Economic Aspects

10

Egyptian Agricultural Policy and the Challenges of the 1990s

Ahmed A. Goueli

The agricultural policy of Egypt has gone through several changes in the past forty years. These changes were incremental in some years, dramatic in others. They reflect the government's role in agriculture and the economic ideology adopted. The economic reform program being implemented at present, including agriculture as a leading sector in the economy, is designed as a comprehensive package—a package consisting of economic and institutional changes.

The implementation of the reform program has so far been successful in correcting the exchange rate, interest rate, control of public expenditure, and to some extent elimination of price distortions. Further, Law 203 of 1991 dealing with the state-owned enterprises, which is now in the process of implementation, is an important institutional reform. Elimination of distortions in the economic system—liberalization of markets—is a necessary but not sufficient condition for achieving the objectives of the economic reform program. Institutional change in the economy, as well as in the agriculture sector, is the more challenging issue facing the decisionmakers in the process of reform, because it is both complex and politically sensitive.

This chapter outlines a framework for agricultural policy in the 1990s to achieve sustainability in the agricultural sector. Sustainable agriculture has several definitions. For my purpose, the definition adopted by the Consultative Group for International Agricultural Research (CGIAR) is satisfactory, i.e., "successful management of resources for agriculture to satisfy changing human needs while maintaining or enhancing the quality of the environment and conserving natural resources" (Walsh 1991). Successful management implies at least the preservation of the same number of available choices in the future. Achievement of sustainable agriculture and its maintenance depends not only on physical and technical factors but perhaps more importantly on economic, social, cultural, and political factors. The agricultural policy designed to achieve sustainability must be comprehensive and integrated, with adequate institutions and services for its implementation.

History of Agricultural Policy

The geography of Egypt and the importance of the River Nile, coupled with the strategic role of agriculture, historically created a tendency towards centralization and strong control of agriculture. Transfers of surplus from agriculture were used to realize the ambitions of the rulers of Egypt. Government developed the irrigation infrastructure and farmers delivered their produce at a very low price. This policy was documented in detail by the French when they occupied Egypt at the end of the eighteenth century. The same policy was followed by Mohammed Ali in the nineteenth century and was largely maintained by the British colonial government.

After the Revolution of 1952, the government's role in agriculture increased even more and several new institutions were created to implement agrarian reforms. The present organization of agriculture is to a large extent the result of the cumulative process of policy changes since 1952. A study of this process is essential for identifying the required policy changes to attain sustainable growth in the agricultural sector. For the purpose of this study, the period of forty years from 1952 to 1992 can be divided into five phases: (1) Nationalism, (2) Socialism, (3) Open-Door Policy, (4) Revitalization, and (5) Economic Reforms. These phases are highly interrelated, but each has different features with respect to the policy instruments.

Nationalism: 1952–1961

The land reform law of 1952 was the new government's first major intervention in agriculture. The importance of the law stems not only from the establishment of a ceiling on agricultural land and distribution of the excess land to the landless, but also from the regulation of tenancy. Agricultural rent was set in the law at seven times the value of agricultural land tax. It gave tenants the rights of inheritance that made them the virtual owners of the lands they would cultivate. Further, the land subject to the new law was organized into cooperatives, known as the "land reform cooperatives," and administered by the village and district offices and a central agency. These regulations and institutions are still an integral part of Egyptian agriculture.

The system of agricultural cooperatives was expanded to include the land not involved in land reform, known as credit land. The government's program for land reclamation also started in this phase, when around 78,000 feddans were reclaimed. The land reclamation policy was based on the ownership of small farms for better utilization of the land resource.

During this period, the cotton trade (domestic and export) was also nationalized and controlled directly by the government.

Socialism: 1961–1974

Dramatic changes in agricultural policy occurred during this phase. Government controls reached their climax. A policy of low agricultural prices to transfer the surplus of the agricultural sector was adopted to finance the urban-industrial sector. Public agencies were created to run most agricultural activities, from farmgate to consumer, both domestic and foreign. Land reclamation and cultivation of the new lands were organized around state farms. The High Aswan Dam (HAD) was constructed. Accordingly, one of the largest programs for land reclamation was implemented, involving 500,000 feddans. The rate of growth in agriculture was as high as 4 percent per year due to the transformation of the basin irrigation land to permanent irrigation and a shift in the sowing seasons of maize and rice crops. Following the June 1967 war with Israel, the country went through a very difficult time: public investment stagnated in the economy in general and in agriculture in particular.

Open Door: 1974–1982

After the October 1973 war, Egypt adopted the "open door" (*infitah*) policy to attract foreign investment. The slogans of "food security" and "green revolution" were raised. The private sector was allowed, in a very unorganized fashion, to reclaim desert land for agricultural use and for speculation. Due to the increased amount of workers' remittances in the mid to late 1970s, urban encroachment on the agricultural lands increased rapidly. But the same policies of the 1960s on depressed agricultural prices prevailed. Due to these policies, coupled with a low rate of investment of less than 8 percent of national investment, a high rate of agricultural growth could not be sustained. The most important institutional change was the creation of the village bank system, which took over all the functions of village credit cooperatives in 1977. The period witnessed an unprecedented increase in the number of loans for food security, extended at a highly subsidized rate (6 percent). This led to a huge expansion in the number of poultry farms. The only positive change in the 1970s was the start of foreign aid (particularly from the United States) for transfer of agricultural technology.

Revitalization: 1982–1990

The food gap in Egypt had widened to an unprecedented level in the 1970s: in 1980 it was ten times the size of 1970. This alarming gap, increasing dependence on the international market, and a stagnant agriculture diverted the policymakers' attention to agriculture. An active policy for the transfer of new agricultural technology was adopted. Gradual relaxation of the state controls was part of the new agricultural policy. Liberalization of agriculture was at the heart of the new strategy. By the end of the 1980s, several results had been realized:

- Compulsory deliveries of crops were eliminated, except for cotton, sugarcane, and rice (the rice quota was abolished in 1991)
- Agricultural prices were allowed to rise and align with world market prices
- The private sector was permitted to import and trade in fertilizers, pesticides, and feeds
- Administered consumer prices for several foods were eliminated and left to the market mechanism (including meat, vegetables, and fruits); further steps were taken to reduce consumer subsidies

In this period, several institutional steps were also taken, including the establishment of the General Agency for Aquaculture and agricultural mechanization stations. The latter were later transformed into a state company. There also was passage of Law 116 (1983) to protect agricultural land from urban encroachment. In the area of land reclamation, the private sector was encouraged. A new program for youth graduates of technical high schools and universities to own and manage small areas of reclaimed land was actively implemented. The declared policy of the government in the new lands was its responsibility for building the infrastructure and the private sector's responsibility for land utilization.

By the end of the 1980s, productivity of major crops—wheat, maize, rice, fruits, and vegetables—increased significantly compared with the 1970s. Credit for the increases went to policies on price liberalization and the transfer of technology, especially the introduction of high-yielding varieties. But cotton productivity declined sharply in the 1980s due to several factors, including government controls.

Economic Reform: 1986–1992

With support from the IMF and the World Bank, in 1991 Egypt began a comprehensive economic reform program, emphasizing trade liberalization and structural adjustment. In agriculture, the process of liberalization and privatization had started in 1986. Agriculture is now the most privatized sector of the Egyptian economy, with private landownership reaching 97 percent of the farming community. With the current economic reform, agriculture has been directly affected in a number of ways:

- The interest rate increased from a low (subsidized) level of 6 percent to a (market) interest rate of 20 percent
- There was active participation of the private sector (besides the Agricultural Bank) in the trade and distribution of agricultural inputs, particularly chemical fertilizers
- State-managed production enterprises were transferred to the new system of holding companies (Law 203 of 1991)
- The state marketing system for all crops except cotton was privatized
- There was limitation of the scope and amount of agricultural import subsidies

Due to these changes and further anticipated adjustments the role of the government in agriculture is witnessing drastic changes.

Major Challenges

Egyptian agriculture faces, at present and in coming years, several major challenges. Agricultural policy has to be reformulated to enable agriculture to achieve societal objectives such as economic growth, poverty alleviation, and sustainability. The challenges can be classified into four interrelated groups: (1) domestic, (2) international, (3) resource base, and (4) technological.

Domestic Challenges

The present high population growth rate (2.7 percent per year) is considered the most important challenge for Egypt. The population of 57 million is expected to rise to 65 million by the end of the next five-year plan, in 1997, and to around 70 million by the year 2000. About one-half of the population lives in rural areas. The high population growth rate is a major constraint for sustainable development in Egypt. It creates a widening food gap, increasing unemployment, and deterioration of the environment. The food gap has increased dramatically since 1974: its value was about US$1.9 billion in 1980, compared with US$184 million in 1970. Self-sufficiency in basic foods declined to 22 percent for wheat, 66 percent for corn, 30 percent for oils, 52 percent for sugar, and 75 percent for meat. Food imports had to be financed through foreign debt and food aid and the trade balance, which was positive up to 1974, deteriorated. Agricultural exports cover only 20 percent of agricultural imports. In the 1980s, the average annual bill for agricultural imports was US$3.6 billion, compared with US$700 million for agricultural exports. Food imports alone averaged around $2.9 billion.

It is expected that food demand will expand but without substantial expansion in domestic food supply. This situation will make it difficult to maintain the nutritional needs of the population, particularly of the low-income groups. A policy of self-reliance, by increasing foreign agricultural earnings, is required through promotion of exports and improvement of productivity of the main food staples. Further, an efficient consumption management policy and effective family planning program are badly needed.

Population growth adds to the already high level of unemployment in Egypt. The unemployment rate at present is between 10 to 15 percent of the labor force. It is expected that eight million persons will be added to the labor force from now to the year 2000. Increased unemployment is a major threat to sustainable development, because its impact is not only economic but also social and political.

Population density is very high in Egypt: according to the 1986 census, about 1,170 persons per km^2. As much as 97 percent of the population is living in only 4 percent of the area. In the rural governorates, density is about 2,512, 1,485, 1,450, and 1,582 persons per km^2, respectively, for Qualyobia, Gharbia, Menofia, and Sohag. For Cairo it is 28,332 per km^2. The high density in the Nile Valley and Delta is responsible for increasing deterioration of the environment. It has transformed the best agricultural land to urban uses, and created a multitude of environmental problems, especially pollution. This situation works against the sustainable development of rural areas. Geographical redistribution of population may have to be done on a large scale. The most effective measures may be the establishment of communities in the new lands and regional planning.

International Challenges

Egypt has succeeded recently in lowering its foreign debt and has signed an agreement with the International Monetary Fund (IMF). Although these are favorable events, a new international situation is emerging, with serious consequences for Egypt and other developing countries. The collapse of the USSR and the liberalization in Eastern Europe has cost Egypt traditional markets, especially outlets for agricultural goods. These countries have themselves become major recipients of food aid and foreign capital. To this must be added the evolution of the EEC and its conditions for market entry: this has further constrained Egyptian exports. For sustainable development, dependence on food aid must be lowered; mobilization of domestic savings is needed; and strengthening of the agricultural export sector is essential. To achieve these goals, Egypt will require a high degree of flexibility in the economy to adjust efficiently and quickly to the dramatic and fast changes in the international marketplace.

The Natural Resource Base

The decade of the 1980s was a period of low water-flow in the River Nile. This low flow increased the society's awareness of water as a scarce resource. With ambitious targets of land reclamation set for the year 2000, the available water resources will be greatly stretched—a situation that raises the question of how to use water efficiently. Irrigation water is treated as a "free good" by farmers as long as they do not directly bear the cost of at least its delivery from the irrigation system. It is essential to introduce a water pricing system to discourage waste and change the existing cropping patterns in which water-intensive crops (sugarcane) are replaced by others to get a better return on investment. The scarcity of irrigation water also leads to the use of low-quality water. The Egyptian water plan includes the use of agricultural drainage water mixed with canal water, treated wastewater, and underground water. Sustainability of productivity in agriculture requires a high degree of care in the use of low-quality water due to its adverse impact on land and crops. Severe measures are also required for the protection of the Nile water from pollution and proper maintenance of the irrigation system. A cost recovery formula is needed to finance these programs.

Technology

Agriculture in Egypt has received technical assistance from several sources in recent years. The impact of these efforts is reflected in the high crop yields and varietal diversity. The main purpose of this technology transfer was the acceleration of growth; it was less concerned with sustainability research. Accordingly, the intensive use of chemical fertilizers and insecticides—the insecticide treadmill—was essential for utilizing the available technology. Moreover, farmers are intensifying the use of agricultural land. The three-year traditional crop rotation has been replaced by a shorter rotation. The system has become unsustainable to a large degree in the new lands. Research must now take the sustainability objective into consideration. The research programs must assign high priority for low-input use for increasing output and productivity. It should also be noted that sustainability research is a long-run undertaking. With the high dependence of the Egyptian agricultural research system on foreign aid, its own sustainability is in jeopardy. This raises the question about the time horizon of foreign assistance and also the creation of a system for increased self-reliance.

Goals and How to Get There

Acceleration of productivity is the main goal of Egyptian agriculture. Increasing the productivity of small farmers will fulfill the national goal of poverty alleviation. But in achieving these goals environmental and resource sustainability should not be sacrificed. The challenge facing the policymaker is: how can agricultural development (1) reduce poverty; (2) be made compatible with maintenance of the natural resource base; and (3) be sustained in the long run? Although strongly linked, the goals of growth, poverty reduction, and sustainable use of natural resources usually are addressed separately by policymakers. To attain these goals—which are not necessarily in conflict—agricultural policy must be designed in a way that reflects the dynamic nature of sustainability. It must be the integrated framework, combining the technical, economic, and environmental aspects.

The administration of this policy requires strong coordination between institutions, from government to the farmer. A critical examination of the existing institutions serving agriculture reveals serious deficiency in coordination. This inadequacy has resulted mainly from pursuing a policy of central control that created institutions with strong vertical links and weak horizontal links. The major challenge facing us at present is how to set up an institutional framework compatible with the new philosophy of a market-oriented economy. The role of government under liberalization is far more difficult than the role under controls. Agricultural policy with controls was relatively easy to implement by using force, although its result was often perverse for promoting sustainable development. In contrast, agricultural policy under liberalization uses indirect instruments and relies on market mechanisms to achieve its goals.

Policy Design and Administration

Egyptian agricultural policy should reflect the country's agricultural situation in the following ways:

- Greater attention must be given to cotton so that it regains its importance as an export crop
- More reliance should be placed on domestic supplies of cereal crops
- Integration of the old and new lands—with regard to the distribution of resources and marketing—should be increased
- Imbalance between crops and livestock should be redressed

Development and Conservation

The water and land resources of Egypt must be improved to maintain a sustainable agriculture by focusing on the following actions:

- Increase the sustainability of water resources in terms of quantity

and quality
- Introduce water into economic accounting for various agricultural uses
- Establish a system of cost recovery to maintain the irrigation system
- Introduce water pricing in the new lands
- Redistribute population among regions to limit the deterioration of environment and urban encroachment on agricultural land
- Use low-input technology to limit the pollution of water, canals, and soils

Institutional Reforms

The past *dirigiste* policy created great dependence of village institutions on government support and management. Village cooperatives and credit institutions need reconstruction. An expanded role of cooperatives and private enterprises is needed in the areas of input trade, marketing, exports, and processing.

Agricultural Legislation

The role of the state is to introduce changes in the existing laws and introduce new laws. At present, some of the agricultural laws are not suitable for economic liberalization. The land reform law was promulgated in 1952 and has to be amended in response to the changing conditions. The law should be examined and changed in the direction of more flexible and fairer rental and tenancy arrangements. Similarly, other laws must be examined, e.g., in the areas of seeds and agricultural quarantine.

Creation of Markets

The process of transition from a controlled to a free economy may create a vacuum in the absence of both state and private enterprises. The state has to support the private sector investment in agricultural marketing. Training in this area will be valuable. The role of the government is to issue regulations that will maintain fair competition and limit monopolistic tendencies. Integration with international markets requires efficient and flexible private and public institutions.

Rural Development

Rural development should be regarded as an important part of strategy for Egypt. Rural roads are important for agricultural marketing and regional integration. Small-scale activities, if created, could absorb a large proportion of the rural labor force. The Social Fund for Development (SFD), as part of the adjustment program, has recently started some of these activities. Likewise vocational training for the rural nonagricultural labor force is essential for the alleviation of poverty. Rural infrastructure should be developed with farmer participation, in which the government acts as no more than a catalyst.

Research and Extension

Agricultural research institutions are dispersed among government departments and universities. Strong ties among these institutions should be established to make them work more effectively. The same problem exists with regard to the extension institutions. The most serious problem in agricultural research is its dependence on the international community for financial support, which makes long-term funding uncertain. These conditions constrain Egyptian research institutions from doing enough sustainability research. The creation of self-reliant finance is important, and the international community supporting the national research effort should be requested to extend the time horizon of its support.

References

Walsh, Jon. "Presenting the Option: Food Productivity and Sustainability," *Issues in Agriculture 2*. Washington, D.C.: CGIAR, 1991, 7.

11

Structural Adjustment and Egyptian Agriculture: Some Preliminary Indications of the Impact of Economic Reforms

Ngozi Okonjo-Iweala and Youssef Fuleihan

In May 1990, the Government of Egypt embarked on a comprehensive structural adjustment program supported by the World Bank. The aim of this program was, and is, to move the economy from one primarily dominated by the public sector, with severe distortions, to a more market-oriented one, by liberalizing prices, removing restrictions on trade, privatizing public sector enterprises, and generally creating a more favorable environment for private sector activity. The program aims to address the main structural weaknesses of the Egyptian economy within the framework of a macroeconomic stabilization effort, supported by the International Monetary Fund (IMF). The underlying impetus of the government's program is the desire to lay the foundation for renewed economic growth in the medium to long term, while at the same time minimizing the negative effects of economic reforms on the poor through improvements in social policies.

The agricultural sector has been at the forefront of these reforms. Starting as early as 1986, prices for most crops were liberalized (except for cotton and sugarcane), crop area allotments were removed, and marketing restrictions and controls for most major crops were also removed. These early reforms were based on analytical work carried out by the Ministry of Agriculture and Land Reform in the early 1980s, which indicated the deleterious impact of the macroeconomic and sectoral distortions on agricultural production and trade. Both the macroeconomic and sectoral reforms are far from complete. Major areas of structural reform remain to be addressed, and it is too early to see the full impact of reforms already undertaken. Nevertheless, as this chapter outlines, there are indications that these reforms are beginning to have some impact on the agricultural sector. We will also set out the agenda for further sectoral reforms.

127

The Prereform Period

Egypt's development strategy from the 1950s onwards was inward-look-ing, emphasizing import substitution and a high level of self-sufficiency, while stressing social welfare objectives. In the centrally planned econ-omy, the public sector played a dominant role in the Egyptian economy, and its contribution to the gross domestic product (GDP) and employ-ment was among the highest for developing countries. Even today, the public sector accounts for over one-half of GDP and 75 percent of manu-facturing value added.

Economic development was financed by large foreign exchange in-flows from donors and external borrowing, local borrowing, oil exports, remittances of workers in the Gulf countries, Suez Canal revenues, tour-ism, and direct foreign investment. The "open door" policy initiated by President Anwar Sadat in 1974 led to a partial liberalization of the economy, which brought with it the seeds of change to a market economy. As a result of these favorable circumstances, the Egyptian economy achieved a high growth rate of 8.5 percent per annum during the 1974–1985 period. Rapid growth, however, was achieved at the expense of mounting fiscal deficits, high inflation, and expansionary monetary policy; it also did not bring about a proportionate increase in employment.

During this period, the centrally planned Egyptian economy was characterized by severe distortions of incentives and misallocation of resources through state monopolies that restricted competition; and by controls on the prices of product and factor markets, foreign exchange movement, and trade. This policy, however, was not sustainable in the long run, leading to mounting fiscal and current account deficits, high inflation rates, and a significant decline in oil-related foreign exchange revenues (domestic oil production and workers' remittances), and in-creased foreign borrowing. This in turn led to rising foreign debt, reduced imports and investment, and a sharp slowdown of economic activity that translated into low rates of growth of GDP at about 2.5 to 3.0 percent in the late 1980s.

The Agricultural Sector

Agriculture has been a major source of economic growth in Egypt. His-torically, agriculture dominated the Egyptian economy. In 1974, it ac-counted for 30 percent of GDP in real terms, 25 percent of exports, and about 47 percent of employment. However, by 1990, its contribution to the economy had declined to 20 percent of GDP and 20 percent of commodity exports. The sector employed about 33 percent of the labor force. This decline was compounded by sectoral distortions and by a declining share of agriculture in public investment, with a less than optimum allocation between agricultural subsectors, including an emphasis on land reclamation. For example, agriculture's share in public investment fell from 7.9 percent of total expenditures during the First Five-Year Plan (1983–1987) to 6.9 percent during the Second Plan (1988–1992). It is, however, expected to rise to 15.8 percent in the Third Plan (1993–1997). At the same time, the proportion of public investment in agriculture allocated for horizontal expansion (increasing the cultivated area through land reclamation) increased from about 40 percent in the First and Second plans to 55 percent in the Third Plan.

The value added by agriculture grew in real terms at average annual rates of 2.7 percent in the 1970s, reflecting mainly the beneficial effects of the High Aswan Dam. Once the gains in cropping intensity made possible by controlled releases from the dam were realized, the major sources of growth in agriculture have been yield increases and shifts in the cropping patterns from lower to higher value crops and livestock, which represent activities not controlled by the state. Due to the increased share of public investment going to land reclamation, the reclaimed land became another source of growth, although it contributed far less than its potential. An estimated 1.9 million feddans have so far been reclaimed, representing around 24 percent of the cultivable land, but they contribute only about 7 percent of the gross value of agricultural production. Despite these gains, the growth rate in agriculture declined to 2.5 percent in the 1980s due to extensive government intervention, implicit taxation of the sector, and low productivity of the reclaimed lands.

The major distortions in agriculture arose from controls on areas, production, and prices of crops (which were set below international market prices); government monopoly in the marketing of major crops and distribution of the main inputs; an overvalued and multiple exchange rate regime that gave rise to antiexport bias and discriminated against the private sector; trade restrictions on the import and/or export of various commodities; subsidies for the main inputs; a ban on the production of certain kinds of agricultural machinery and equipment; and highly negative interest rates in real terms. Another major distortion was the provision of irrigation water and drainage facilities virtually free of charge to farmers, hence the absence of incentives for optimizing returns per unit of water used and ensuring the sustainability of the system in an economy short on water and land resources.

Egypt has followed an inward-looking, protectionist trade policy that promoted import-substitution to attain higher levels of self-sufficiency. The exports and imports of several commodities were banned, or subject to quotas or prior approval. This included citrus exports to the major markets. The consumption of some commodities was subsidized, such as wheat, corn, sugar, and vegetable oils, and this contributed to a widening trade gap. Import bans/restrictions were imposed on poultry and other meats primarily for the protection of local industry. Government intervention resulted in highly negative effective rates of protection (ERPs) in the 1964–1985 period for cotton and rice (–56 to –78 percent) as well as for wheat and maize (–24 to –63 percent) (Dethier 1989). These figures show that price controls discriminated against the main crops through 1985. The impact of implicit taxation of agriculture in 1985, excluding irrigation and drainage systems, has been a net transfer out of the sector of over £E 5.5 billion in 1991 prices (£E 2.2 billion in 1985 prices); over one-half of this amount was generated by one commodity alone, cotton (Deither 1989 and Baffes 1992). These transfers, in the case of cotton, resulted in both area and yield decreases of 13 percent and 6 percent, respectively, during 1980–1985.

Trends in areas, yields, production, and exports of the major crops (cotton, wheat, rice, corn, and berseem), which account for 70 to 80 percent of crop area since the 1970s, provide a picture of sectoral performance prior to the reforms (Table 11.1). The highest increases in area, yield, and production in the 1970s were in uncontrolled horticultural crops and livestock, which provided much of the sectoral growth. The area put to berseem rose rapidly in the 1960s and 1970s on account of protection and input subsidies given to the livestock subsector; high population growth, rising income, and high income elasticities contributed to the increased demand for these major commodities. Further, the yields of wheat and corn rose throughout the 1970–1985 period due to the adoption of high-yielding varieties.

The Reform Program (ERSAP)

The unsustainability of macroeconomic measures led to the adoption in 1990 of a bold and comprehensive structural adjustment program that aims to exploit (1) the advantages of the Egyptian economy as a large domestic market; (2) the favorable climate; (3) perennial irrigation; (4) proximity to major overseas markets; (5) a diversified industrial base; (6) relatively skilled and cheap labor; and (7) the tourism potential. This program has been supported by financial assistance from the donor community, including generous debt relief.

The major objective of ERSAP—the Economic Reform and Structural Adjustment Program—is sustainable economic growth, taking into consideration the social costs of structural adjustment. ERSAP aims at restoring macroeconomic balance and structural adjustment of a major segment of the Egyptian economy to improve resource allocation efficiency and move toward market determination of production and trade. The impact of structural adjustment on the poor is being addressed through a social safety net, supported by the Social Fund for Development

Table 11.1 Indices of Cropped Area, Production, and Yield (1980=100)

Crop	Area		Yield		Production	
	1985	1990	1985	1990	1985	1990
Sugarcane	99	108	110	119	109	129
Gardens	127	183	105	102	133	186
Vegetables	105	107	140	144	147	154
Wheat	89	147	117	162	105	239
Long Berseem	112	96				
Short Berseem	93	80				
Beans	123	125	122	142	150	178
Corn	100	104	114	144	115	149
Rice	95	107	100	116	96	124
Cotton	87	80	94	73	82	58
Sorghum	83	78	103	127	85	99
Potatoes	106	113	115	119	122	135

Source: CAPMAS, *Statistical Yearbook.*

(SFD), to mitigate the effects during the transitional adjustment period. Specifically, in the first stage, ERSAP aims at:

1. Reducing inflation and the current account and budget deficits, unification and liberalization of the exchange rate, and restoring credit worthiness
2. Privatization and/or restructuring of public enterprises
3. Financial sector reforms to liberalize interest rates, improve the solvency of banks, and strengthen banking regulations and supervision
4. Liberalization of most prices in agriculture and manufacturing, and long-term marginal costing or international pricing of transport and energy prices
5. Trade liberalization

Significant reforms have already been introduced by the government within the context of ERSAP. These include: (1) raising energy, transport, and cotton prices; (2) liberalization and unification of the foreign exchange market; (3) liberalization of interest rates, abolishing interest ceilings and credit controls, recapitalization of public banks, and organization of a treasury bill market; (4) introduction of a new sales tax; (5) abolishing investment licensing requirements except for a short negative list; and (6) approval of a new public enterprise law (Law 203), the establishment of the Public Enterprise Office, and initiation of the privatization program. However, much more remains to be done, particularly in the difficult area of privatization where the pace is slow and sometimes faltering.

Reform Program in Agriculture

It was widely accepted in Egypt by the mid-1980s that the major constraints to agricultural growth were policy based. In 1986, and within the framework of the strategy for the agricultural sector for the 1980s, the Ministry of Agriculture and Land Reclamation launched a reform program aimed at liberalizing and transforming the agricultural sector to one basically responsive to market forces, taking social aspects into consideration. The central objective of the reforms is to increase production and income through efficient resource use. To attain that objective, several measures have been implemented since 1986:

- Crop area allotments with delivery quotas have been removed for all major crops except cotton and sugarcane.
- Agricultural producer prices for all products, except cotton and sugarcane, have been liberalized; cotton procurement prices have been raised to two-thirds of international prices; and subsidies on fertilizers and pesticides have been significantly reduced, and are expected to be completely phased out by the end of 1993.
- Privatization of reclaimed public lands is progressing, and a program for the divestiture of the assets of sixteen production companies which own eighty-three projects has been initiated.
- Interest rates have moved closer to commercial banking levels
- Trade and marketing restrictions are being relaxed, including the monopoly of the Principal Bank for Development and Agriculture Credit (PBDAC) on input distribution.

Impact of the Program

The period that has elapsed since the start of ERSAP and the reform program in agriculture is not long enough to make definitive conclusions regarding the impact on the agricultural sector. This analysis is preliminary and illustrative.

The impact of ERSAP and the generous debt relief provided by the Paris Club countries have led to an improved macroeconomic performance that is better than was anticipated when the Structural Adjustment Loan (SAL) was approved in 1991. GDP was estimated to have risen by 2.0 percent in the fiscal year 1991, compared with a projected decline of 1.5 percent; this is mainly attributed to higher than projected growth in tourism, workers' remittances, Suez Canal revenue, and construction. The

20 percent inflation rate for fiscal year 1991 was below the 30 percent forecast rate due to reductions in the fiscal deficit and exchange rate stability.

The balance of payments is much better than projected earlier, mainly due to the reduction in outstanding debt to the Paris Club countries, which decreased from US$51.1 billion in 1991 to US$39.0 billion in 1992; this saves Egypt annual payments of US$1.5 billion for interest and amortization. The current account balance shifted from a deficit of about US$2.6 billion in fiscal year 1990 to a surplus of US$2.4 billion in 1991, mainly due to official grants, with a projected surplus of US$1.2 billion in 1992; and international reserves were expected to exceed US$9.0 billion by end of fiscal year 1992 (sufficient to finance 6.8 months of imports).

The fiscal deficit is projected to fall to 7 percent of GDP, as against an initial estimate of 9 percent; this is largely due to less expansionary fiscal policy (except subsidies), reduced interest payments following the Paris Club debt relief agreement, and to some extent the introduction of the 5 percent sales tax. Interest rates on new treasury bills have stabilized around 20 percent, and have caused interest rates of commercial banks to rise (although they are still negative in real terms on an ex post basis). This has encouraged capital inflows and discouraged depreciation of the Egyptian currency.

Despite some delays, most aspects of Egypt's stabilization and adjustment program are being implemented; however, many reforms are still needed in the areas of trade, fiscal, and financial reforms, as well as restructuring and privatization of public enterprises. As a result of the favorable circumstances arising from debt relief and better than expected performance of the Egyptian economy, Egypt faces a unique opportunity to accelerate the pace of reforms, particularly restructuring and privatization, in order to encourage greater private investment and participation in the economy, and achieve the goal of sustainable growth. Currently, Egypt has a strong balance of payments position; the danger exists, however, that the same level of financial support from the international community, if needed, may not be available in the future due to competing demands on donor assistance.

Impact on Agriculture

Macroeconomic and sectoral reforms introduced so far have reduced distortions and preliminary data indicate that there has been some improvement in efficiency of resource use in the sector leading to an increase in production of some crops and a decrease in the output of others. Changes in technology—notably yield increases—have also contributed to the increases in output. Further work needs to be done to disentangle the effects of one from the other. Moreover, certain aspects of reforms, for example, interest rate policy reform, have had a short-term unfavorable impact. Increases in interest rates appear to have temporarily dampened the demand for credit for the purchase of agricultural machinery and some other inputs. Nevertheless, overall qualitative judgment of the impact of the reform program based on the available data indicates a relatively favorable outcome for the agricultural sector which should be accentuated with the completion of these reforms.

On the macro side, the effect of reduced distortions in agriculture is reflected by the sharp decline in net transfers from the sector from £E 5.5 billion (in 1991 prices) in fiscal year 1985 to £E 1.1 billion in 1991. Also, the liberalization of area and price controls of the major crops (except cotton and rice, which were liberalized in 1991) has resulted in a considerable drop (30–68 percent and in some instances virtual elimination) in the level of negative effective protection for these crops—except for cotton, where the level of negative effective protection was still high, –30 percent in 1991 (World Bank 1992). Another indication of the continuing reduction in distortions is the fall in financial input subsidies. The £E 250 million maize subsidy in 1985 has been removed, thus reducing distortions in feed prices. Fertilizer financial subsidies have declined dramatically from £E 175 million in fiscal year 1990 to £E 61 million in 1992. These subsidies should be completely eliminated by 1993. After rising to £E 218 million in fiscal year 1990, credit subsidies have plummeted to £E 61 million in 1992 due to substantial interest rate adjustments (estimates from PBDAC, Price Stabilzation Fund, GASC, and World Bank).

The reduction in subsidies should have a favorable impact on the budget thereby partially offsetting the revenue losses from a decrease in the implicit taxation of agriculture noted above. With a further reduction of the subsidies and the completion of reforms, which would permit a fuller supply response from the agricultural sector, the net impact on the budget should be considerably more favorable. This would be helped by improved direct taxation of agriculture. The present land tax, at an average of less than £E 20/feddan per annum, is quite low and needs to be revised. A review of the land tax should take place in the context of a review of the existing tenancy law. This law, by fixing land rent at seven times the land tax and giving tenants and heirs the right to use the rental land in perpetuity, creates disincentives to a more efficient use of land. The government is reviewing the tenancy law with a view to modifying existing legislation to bring land rent closer to its market value. This will also eventually permit a better land tax.

With regard to credit, as a result of declining subsidies to the Principal Bank for Development and Agricultural Credit and credit ceilings imposed by the Central Bank, interest rates in agriculture have risen to about 16 to 20 percent for farmers, and 23 percent for traders. While still negative in real terms, these interest rates appear to be having a short-term dampening effect on the demand for credit. PBDAC reports a drop in the demand for credit for purchases of agricultural machinery and other equipment. In the near term, as markets adjust, interest rate reform should result in a more efficient allocation of scarce capital resources in an environment of financial sector reforms. The projected drop in inflation by government (to around 15 percent in late 1992) and the concomitant fall in the level of nominal and real interest rates should also ease the credit situation.

At the micro level, there are indications that the liberalization of area controls, cropping patterns, and prices, combined with technological changes and decreases in the cost of labor, have led to significant changes in the area devoted to some crops and to increases in net farm revenue for some cropping rotations. While the area devoted to most major crops (Table 11.1), including wheat, corn, and rice, increased between 1985 and 1990, the clearest illustrations of the shift in cropping patterns relate to wheat and berseem. The area used for wheat increased by a dramatic 65 percent between 1985 and 1990, and yields increased by 38 percent during the same period. This, coupled with the increases in the real price of wheat, has led to greater profitability of the wheat crop. Farmers received an average price of £E 26.31 per ardab in real terms during 1987 to 1990 compared to £E 22 per ardab during 1970 to 1986. Sharper increases in the price of wheat failed to occur because of a concurrent sharp decline in world prices of most commodities including wheat during this period. Increases in domestic price levels just served to bring wheat prices up to world market levels. At the same time, there was a significant decrease in the real wage rate from £E 4 to about £E 2 per day (wages typically make up 30 to 50 percent of production costs). These trends in prices, yields, and costs served to make rotations including wheat relatively more profitable than hitherto.

In contrast to the area devoted to wheat, that devoted to berseem declined by about 14 percent from 1985 to 1990. Berseem is grown by many small farmers who typically have small numbers of livestock as an integral part of their farming operation. These farmers will continue to grow a certain amount of berseem for livestock feed. However, berseem is also cultivated by other farmers for commercial sale as a complement to imported livestock feed. Prior to reform, cultivation of berseem had been profitable due to the relative profitability of red meat compared to other products. Unlike the prices for other products, red meat prices were not controlled and imported livestock feed was at the same time sold at subsidized prices. Berseem as an input into red meat production was therefore very attractive. With the freeing of prices of grains, the relative profitability of red meat and thus berseem has decreased, leading to a decline in total area planted.

Real net returns per feddan for five major rotations—short berseem and cotton, wheat and corn, wheat and rice, long berseem and corn, and sugarcane—for the last twenty years are plotted in Figure 11.1. Returns for the two rotations with wheat were flat and relatively low until 1984 when they began to rise, reaching the most profitable position in 1989. Long berseem and corn has been the most profitable rotation for almost the entire period until 1988. Profitability of the cotton rotation peaked in 1979, declining sharply until 1984. While returns for the cotton rotation have been steadily recovering since then, they are the lowest of the five rotations examined to date and will remain relatively low unless price and marketing reforms of the subsector are completed.

The increases in area planted and yields of the major crops have led to increased domestic output, thus permitting some decrease in imports of certain goods. Imports of wheat and wheat flour declined from 7.1 million tons in 1987 to about 6.6 million tons in 1990. Corn imports, however, remained largely unchanged at slightly less than 2 million tons despite a growing population. There is no doubt that Egyptian farmers appear to be responding rationally to the reforms. Thus, the expected response of agricultural suppliers to the economic reform program should show greater increases, provided the reforms are sustained and carried to their logical conclusion.

Figure 11.1 Net Returns to Main Rotations

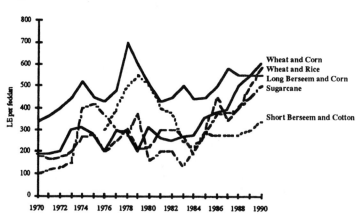

Source: Ministry of Agriculture and Land Reclamation (MALR), USAID Agriculture Database.

Remaining Agenda: Priority Issues

The government has made steady progress on agricultural reforms, and these efforts are beginning to bear fruit. However, there is a need to complete reforms that have already begun in certain areas and to initiate further reforms in others. The government has thus far raised the price of cotton to 66 percent of its border price, thereby reducing one of the most significant distortions. However, the full liberalization of cotton production, marketing, and trade is a high priority given the significance of cotton in the foreign exchange earnings. It is also important to allow the importation of pest-free cotton (subject to strict quarantine regulations) to enable Egypt to export an increasing volume of high-value Extra Long Staple (ELS) and Long Staple (LS) cottons and import cheaper varieties for the production of low-count/value yarn which dominates the production of local mills. A modest tax (not to exceed 10 percent) could be levied on ELS cotton exports to exploit Egypt's position in the world market as a major supplier of extra fine cotton. The government should also facilitate the establishment of a cotton exchange to aid marketing.

With regard to sugarcane, the production of which has been encouraged through the payment of prices above international market levels, the government should fully liberalize its production, marketing, and prices while simultaneously seeking the substitution of sugarcane by sugarbeet, or other less water-intensive crops. Sugar mills should be rehabilitated to facilitate the processing of sugarbeet; and the retraining and redeployment of any labor affected by the reforms should be ensured through the development of an appropriate safety net via the Social Fund and other means.

A program is already in place to ensure the complete phase-down of fertilizer and pesticide subsidies by 1996. The program should be pursued and completed as planned. In addition, the program to privatize existing public sector production of fertilizer and pesticides, which has also begun, needs to be pursued in a systematic manner.

The most important area in which distortions are evident but reforms yet to be initiated is that of *water charges*. Egyptian farmers presently pay very little for water delivered to them through the extensive irrigation system and for drainage work. The only charge farmers are required to pay covers the capital costs of the tile drainage, but these payments are modest since they are spread over a twenty-year period without interest. Given the competing demands on the central budget, and the falling share of agriculture in investment expenditures except for those earmarked for horizontal expansion in the Third Plan, there is an urgent need to recover, at the minimum, the operation and maintenance (O&M) costs of these facilities, and preferably, part of the investment cost.

Water charges are critical not only for ensuring the long-term sus-

Figure 11.2 Water Consumption, Land Use, and Value Added of Major Crops

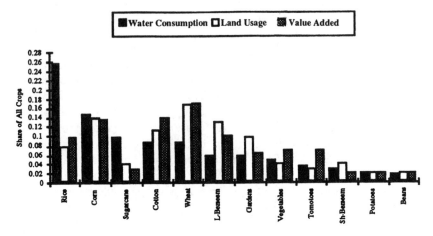

tainability of the irrigation and drainage infrastructure, but also to provide an incentive for a more efficient allocation of scarce water supplies. With the progress of liberalization in the sector, farmers are increasingly free to make rational choices regarding crops and rotations. In the absence of water charges, farmers would use irrigation water until private marginal benefit from irrigation is equalized with private marginal cost, which is much smaller than marginal social cost; therefore farmers have very limited incentive to improve on-farm water-use efficiency. This problem is illustrated in Figure 11.2.

It is clear that sugarcane and rice make a high demand on water resources in comparison to the value added. Preliminary estimates indicate that these two crops consume about 37 percent of total irrigation water, but account for only 12 percent of the cropped area and 13 percent of the agricultural value added. Unless measures are instituted to increase the price of water—and barring the use of area controls, which are more cumbersome and have just been dropped—liberalization could have a perverse effect through increasing the cropped area of water-intensive crops. While there may be merit to cultivating a certain area of rice, because of its use as a reclamation crop, the intensive use of water by rice and sugarcane in particular merits serious government review and action. The government has shown its willingness to consider the issue through instituting a study of the costs of the irrigation system with the assistance of USAID. It is expected that this study will also provide options for improving cost recovery, which will enable the government to act in an appropriate way on this important economic, yet delicate social and political, area.

References

Baffes, J. "Egypt—The Impact of Structural Adjustment on Agriculture." Washington, D.C.: World Bank, April 1992.

Dethier, J.J. *Trade, Exchange Rate, and Agricultural Pricing Policies in Egypt.* Vol. 1. World Bank Comparative Studies Report, Washington, D.C., 1989.

World Bank. *Arab Republic of Egypt: An Agricultural Strategy for the 1990s.* Draft Report, Annex 7, "Economics of Crop and Livestock Production." Washington, D.C., 1992.

12

The Agricultural Sector in the Context of Egypt's Structural Adjustment Program

Salah El-Serafy

The welcome attention to sustainability indicates concern for the survival of the Egyptian economy during the 1990s and beyond. Given this long-term perspective, we have the freedom to be imaginative and think in terms of radical changes, not of drifting from year to year, attempting to cope with everpresent crisis. That the situation is of crisis proportions cannot be denied. Population growth has eroded the land-labor ratio that gave Egypt its past enormous advantage. And water has become progressively scarcer.

Patrick O'Brian is quoted by Bent Hansen (1991) as estimating that while in the nineteenth century population rose fourfold to about 10 million, agricultural production, following major irrigation and crop-pattern reforms, probably rose tenfold. Thanks to agricultural growth, the standard of living more than doubled during the present century. However, in this same period population has grown five to sixfold; and agriculture, confined to a limited area with increasing signs of water scarcity, has been unable to deliver the requirements of the population's rising expectations. Even what appeared at the time to be drastic—though intermittent—measures, such as the heightening of the old Aswan dam, the construction of the High Aswan Dam, land redistribution, and the agrarian changes that followed the 1952 Revolution, have turned out in retrospect to be no more than palliatives.

Thus, despite many efforts, agriculture receded in relative importance in the national economy. In fact, even if we count as agricultural those products with agricultural content, such as textiles, we must conclude that the sector is not contributing much to the balance of payments. Egypt imports vast amounts of grains for food and fodder; and the agricultural sector also imports many implements, and much pesticides and fertilizer; it also absorbs exportables, such as fuel and urea, which would otherwise have contributed to exports. As population continues to grow at a frightening pace, without major changes in the economy and the agricultural sector, agriculture will continue to show diminishing returns and be of declining relative importance, failing to contribute to the recovery of Egypt's economy.

A Disastrous Record?

Taking the long view, and based on computations of total factor productivity, Hansen (1991) has concluded that the data "reveal how disastrous Egyptian agricultural developments in this century have been. With total inputs increasing at an annual rate of 1.3 percent from 1900 to 1967, total output increased by only 0.9 percent and total factor productivity thus declined by 0.4 percent annually." Hansen further estimates a negligible increase in total factor productivity of 0.1 percent per annum in the whole period from 1900 to 1982. Bringing his speculation up to date, he judges that total factor productivity in Egyptian agriculture "may have been about zero over the century." This, it should be stressed, has been the century of major technological improvements in agricultural systems the world over, so that the considerable comparative advantage the sector had enjoyed previously inevitably eroded vis-à-vis outside competitors.

Failure of Industry

Egypt's industry, which could have developed in the direction of light manufactures absorbing the surplus labor, using technology suitable to emerging countries, has unfortunately had an even worse record. For example, between 1954 and 1981, the ratio of domestic resource cost to the exchange rate was well in excess of unity for cotton yarn exported. This ratio was considerably higher for phosphatic fertilizer, paper, iron, and steel. For twelve public sector industries, using shadow exchange rates, high negative values were calculated for "miscellaneous chemicals," basic metals, transport equipment, and china and glass, indicating that such activities impoverished rather than enriched the economy (Hansen 1991).

We have only to look at the balance of payments data to note the heavy demand industry places on imports, and the almost negligible contribution manufacturing makes to exports after a generation or more of attempted industrial development.

Over the past two decades the survival of the Egyptian economy has depended on fortuitous endowments and favorable exogenous factors—which, however, inhibited the development of a proactive economic policy that might have put the economy on the right track. These included the oil price explosions of 1973/74 and 1979/80; location of rich but limited petroleum deposits; migration of Egyptian labor to the oil-rich countries, which has brought in valuable remittances and own-financed imports; oil-related traffic earnings of the Suez Canal—and of course tourism.

Tourism, unfortunately, is not much analyzed from the point of view of its net returns to the domestic economy, nor its net contribution to the balance of payments. It is not my wish to go into the economics of industry and the history of industrial development, such as it has been. I will say only that the pattern imposed on that policy sector (with its public sector organization, inward-oriented, import-substitution policies) and the almost total neglect of export possibilities, and excessive reliance on engineering considerations with scarce attention to economic criteria—these factors lie at the root of the failure of industry to play its proper role. Progress in certain areas of manufacturing can indeed be cited. What is important, though, is not just to record some fragmenting progress, but to compare the overall achievements of Egypt with those of other countries that began to industrialize a generation ago from levels of development comparable with that of Egypt. Had industry filled its proper role, it might have reflected favorably on agriculture and the Egyptian economy as a whole.

The Structural Adjustment Program

Egypt found itself in the latter half of the 1980s unable to pay fully for its imports or service its external debt. After a halfhearted and later aborted attempt at structural adjustment in 1987, Egypt finally and boldly adopted, in 1991, a comprehensive program for economic reforms. Following approval by the International Monetary Fund (IMF) of a stand-by arrangement, favorable rescheduling by the Paris Club of a sizable part of Egypt's external debt obligations, and significant bilateral debt forgiveness, the World Bank was able to come forward with support for an adjustment program after many years of difficult discussions with the Egyptian authorities. A special "Egyptian Social Fund for Development" (SFD) was created as an integral part of the adjustment program, with the objective of alleviating some of the hardships associated with the adjustment. This has been built around a contribution from the International Development Association (IDA), the World Bank's soft-loan affiliate.

Prerequisites for Success

Before the World Bank supports an adjustment program, it usually requires that the borrower develop a medium-term macroeconomic framework that shows how the basic aggregates are expected to change, how the imbalances will be reduced, and the implications, internal and external, of the adjustment process for the economy. In 1991/92, the resource balance, i.e., the difference between exports and imports of goods and nonfactor services, was estimated at a negative 17 percent of GDP; the current account balance a negative 10 percent; and external debt interest payments 10 percent, or about $3 billion (World Bank Report 1991). Under the adjustment program the exchange rate is projected to worsen gradually to 5.6 Egyptian pounds to the US dollar by fiscal year 2000, with the underlying assumption that strong inflationary pressures are expected to build up during adjustment. Adding up the projected deficits on current account for the period from 1992 to 2000, the sum indicates a gross increase in external debt obligations of nearly $30 billion ($18 billion net); as stated earlier, this signals the danger that debt service may once more become a problem. Another alarming trend underlying the program is the projected erosion of per capita consumption, expected to occur for several years in the future, with its level hardly rising by the end of the century above what it had been in fiscal year 1984.

I mention the above not to be alarmist or discouraging, but to emphasize the gravity of Egypt's economic situation and the ambitiousness of its adjustment program. I want to stress that the economic situation requires not just marginal changes, but drastic and radical solutions. It is worth noting that the current World Bank loan for structural adjustment, with its attendant conditionality, if judged to be successful, will be only the beginning of a series of such loans, which are bound to bring in more conditionalities.

Conditionality of Egypt's Loan

Along with other adjustment programs, the first Structural Adjustment Loan (SAL) to Egypt, which was fully disbursed by the end of 1992, was made subject to certain conditions. These conditions aim at ensuring that the loan will be *productive,* in the sense of improving the efficiency of the economy. Experience with adjustment lending has taught that for an adjustment program to succeed, several factors must be present, three of which are paramount. One is the seriousness of the government's resolve to implement the usually painful adjustment program; another is the prospect of enough external financing materializing, on satisfactory terms, to support the program; and a third is the need initially to go through a stabilization phase, to pave the way for successful structural reform.

Despite appearances to the contrary, the movement from the public to the private sector usually supported under adjustment programs, is not necessarily doctrinaire. What it seeks to do is to raise the degree of competition in the system in order to induce greater efficiency. The government's role in the economy will still remain large and important, but the thrust of the change is to take away from the public sector those functions that are not best undertaken by governments in order to enhance the role which only governments *can* play. Such a role would include defense and security, nurturing and sustaining an effective legal system, developing and maintaining the infrastructure, including the environment, and engendering and protecting the forces of competition. It would also cover macroeconomic management of the economy, including poverty-alleviating income redistribution measures, largely through fiscal policy.

The current SAL contains six major areas of conditionality: an agreed macroeconomic framework; privatization; price adjustments; trade liberalization; deregulation; and the Social Fund. For the agricultural sector, conditionality centers on price liberalization, deregulation, raising cotton prices received by the farmer and deregulating cotton marketing and trade—including ginning and exports. It also covers reducing and ultimately phasing out subsidies on fertilizers and pesticides.

Conditionality and Agriculture

The agricultural sector is seen to be an integral part of the general pattern of the SAL conditionality. Egypt's agriculture has been largely in private hands, and many claims have been made that it had escaped being nationalized. Such claims are not convincing in view of the pervasive controls imposed by the government on the sector. Regulations specified which areas would be under certain crops; there was administered pricing for inputs and outputs and control and direction of rural credit. Above all, there was public management of irrigation water and drainage—not to mention extensive controls on imports and exports, consumer prices, and investment licensing, all of which bear heavily on agricultural activities. Just moving prices, trade, and production in the direction of liberalization will not guarantee that Egypt ends up with a liberalized outcome. The spirit of liberalization, however, has been raised, and this is bound to affect agriculture, as other sectors, provided the adjustment momentum is sustained.

The attempt to make agricultural prices, input and output, converge towards international prices is closely related to the major objective of integrating the sector in the world market. Advocates of such convergence often do not realize that while on the face of it convergence indicates efficiency in resource allocation, with potential benefits from trade, a closer look reveals other considerations. Though the balance of the empirical evidence supports the drive to remove input subsidies, such removal might discourage the use of important inputs, and end up in reduced production. Exposing domestic producers to international market prices can bring in price and income fluctuations, as well as the end-results of agricultural dumping by foreign trading partners to the detriment of the sector. Another consideration in the attempt to impose international prices on Egypt's agriculture is the fact that the government already incurs great costs for providing irrigation water (including capital depreciation and recurrent costs) for which it is not compensated.

Agricultural income, it should be noted, is not subject to tax, and many of the output price controls were a form of agricultural taxation. Such taxation tended to be excessive, distorted market signals, and dampened incentives. However, to complete the liberalization drive as reflected in the SAL conditionality, water should also be priced to reflect its scarcity and the cost of proper drainage. Improvements should be attempted in the landowner-tenant relationship, and also in the labor market. Mechanisms should be established, however, to protect the farmer from the excessive price fluctuations that external trade is bound to engender, and to counteract the dumping in the Egyptian market of imports subsidized by foreign exporters.

Adjustment and Sustainability

Considerations of sustainability force us to take a long view. The balance of labor (population), capital, land, and water should be viewed as an integrated input package for Egypt's agriculture. But a question needs to be asked: can the sector continue at a level of acceptable productivity and make an improved contribution to the economy? The sector itself may expand, remain constant, or decline, but how best can we optimize its economic contribution? I have no hesitation in venturing the judgment that the sector can make a substantial contribution and move into a sustainable direction, provided certain changes are made with imagination and resolve.

Note that the structural adjustment per se may not take us along a sustainable growth path. To balance foreign deficits under adjustment programs, countries can lose their environmental base at an alarming pace, and liquidate their natural capital. They may be judged, in retrospect, to have successfully adjusted. But the criteria of balance, to be restored through adjustment, as used for instance by the IMF, are short-term criteria, which may or may not be consistent with long-term sustainability. Often they are not. That is why even if Egypt went successfully through four or five adjustment loans, taking us to the year 2000, there is no guarantee that sustainable development will be secured. The achievement of sustainability requires a special effort and a specific orientation.

Sustainability and the Future of Agriculture

The notion of sustainability has gained currency in recent years. Thanks to the 1987 UN World Commission on Environment and Development—the Brundtland Commission—the concept has now been firmly associated with *environmental* sustainability (World Commission Report 1987). I would like to distinguish environmental (in the sense of preserving for future generations the physical and biological legacy we inherit from the past) from *economic* sustainability, which revolves around maintaining over the long term an undiminished level of income, or more precisely, consumption.

The two views of sustainability overlap, but economists tend to tolerate some decline in environmental capital, provided alternatives are found to substitute for the lost elements to sustain future income. However, economists forfeit all credibility if they lose sight of the necessity for an economy to sustain its prosperity over the long run; if they focus only on the short term while the natural base of the economy is being eroded. The key to economic sustainability lies in the need to repair, maintain, and restore all forms of capital, including environmental capital, as a means of creating current and future income. If capital is not kept intact in the accounting sense, then ostensible income is overestimated, leading to unwarrantedly high consumption—and this is a recipe for certain future decline. Before true income is reckoned, it is imperative that environmental damage has been properly assessed and accounted for in the estimation of income.

In Egypt, pressure of population on limited land and water resources has gone hand in hand with diminishing returns and increased fragmentation of land holdings, and increasing levels of inputs have been used to maintain yields. Agricultural labor, which increasingly became redundant and less productive, migrated to the cities and, with the 1970s petroleum boom, to neighboring countries. This provided some relief to agriculture but it created serious problems for cities and towns. With mounting population pressure on urban centers, valuable agricultural land was turned to urban development, which lessened the pressure on water resources but intensified land shortage. The majority of Egyptians now live in urban centers that are overcrowded, polluted with vehicular and industrial emissions, and served by inadequate and unmaintained sewerage systems. They suffer frequent breakdowns of electricity and water services.

Meanwhile, population growth continues to erode the foundations of Egypt's economic sustainability, diverting attention from long-term development to short-term crisis-fighting, which is undertaken at practically any cost, including environmental cost. The problem is compounded by the fact that a majority of the population is still illiterate—a telling indication of the failure of the educational system allegedly free and universal since 1952. Much the same can be said of health and other public services, which are poor, short of resources, and highly inadequate.

A hard look at Egypt's resources and the growing needs of its rising population raises questions about the ability of the Egyptian economy to maintain even the present low level of per capita income, estimated at some $600 in 1991. Not only are there limited prospects for the leading export—petroleum—but there are doubts about the sustainability of workers' remittances, Suez Canal earnings, and foreign assistance. Such sources of foreign exchange should be pursued while they last, but for sustainability we must focus our attention on *domestic* resources, including Egypt's longer-term comparative advantage in having large supplies of labor. These might be directed to gainful and sustainable employment.

What seems to be certain is that population will continue to grow during the 1990s and for a considerable period afterwards, even with the most optimistic estimates for a decline in fertility consequent upon the adoption of immediate, effective, and imaginative demographic policies. With the inevitable population increase, the needs for food, housing, education, health, and other services will continue to grow, eating away at the economy's ability to save and invest for a better future. Besides, despite the recent decline in foreign debt consequent upon forgiveness and rescheduling, debt and its service are bound to grow again, particularly in view of the large borrowing that is being undertaken and will have to be undertaken in support of structural adjustment.

Major Issues for Agriculture

Since the adjustment effort involves major restructuring of the economy, it will not be fitting, when we consider agricultural adjustment (perhaps the most promising sector for the economy's salvation), to be timid and unimaginative. If, in economic areas, the legacy of government intervention can be discarded in favor of market forces, we should not shrink from contemplating radical and large-scale shifts in the organization of agricultural production. The opening up of Egypt's economy to external trade, as a policy objective, has already been initiated. The price adjustments already made, and others that will follow, together with deregulation and liberalization, indicate an irreversible trend of radical change. "Business as usual," in agriculture or any other sector, will not do.

Factor scarcity: Land, water, labor, and capital are needed to make any agricultural system work. For Egypt, both land and water are scarce and growing scarcer. Whether land or water is the scarcer is moot, with general leanings in the direction of water. For policy purposes, however, we do not have to choose between them. They are both short, and to some extent as limiting factors mutually substitutable. As to labor, formerly it was abundant in agriculture. Though it showed intermittent signs of scarcity, especially seasonally, at present it has such a low productivity that any scarcity it may develop could in theory at least be turned into relative abundance. Its productivity can be raised through various means, including better management and capital investment, better extension services, and improved rural skills.

As mentioned earlier, land scarcity has recently been reinforced by the scarcity of water, 95 percent of which derives from the Nile. Any expansion of agricultural land puts added pressure on water availability, so water may tentatively be viewed as the limiting factor on agricultural expansion. As part of structural adjustment, water must be priced rationally for all its users, in harmony with the trend in other prices. This would limit its use in the aggregate, and direct it to those uses that are most valuable to society. Many practical barriers, however, stand in the way of pricing water, especially water for irrigation. Such water consumption is difficult to measure accurately, and there is strong societal resistance to charging for water. Until a system of water pricing is in place, proper shadow-pricing of water needs to be developed and used to guide the allocation of water and indicate its best use in terms of choice of crops. I will outline a simple approach.

Crop choice: A calculation: In this exercise I shall not ignore the scarcity of land: it is best to proceed with water and land in tandem. On the basis of estimated land and water requirements per crop in the 1990/91 season, and comparing these with the value added by each crop, it is possible to compute indices of land and water productivity separately with respect to value added. Caution should be exercised, since all the basic data must be to some extent suspect: value added relies on approximations of inputs and outputs and their values, and such values are often administered, subsidized, or subject to market imperfections and public intervention. Besides, water and land requirements are also subject to uncertain measurements.

I have estimated the value-added productivities of crops with regard to land and water (Table 12.1). Then I have ranked these crops (Table 12.2) in a descending order (from top to bottom) of their productivity in terms of water and land separately (columns 1 and 2). The same ranking can be read as indicating the ascending order of water and land requirements by each crop. Here we see that the most productive use of land seems to be in tomatoes, potatoes, vegetables, cotton, rice, and beans; and the least productive is in short berseem, orchards, sugarcane, and other crops. Wheat and corn fall in the middle of these two extremes.

As to water, it seems that tomatoes, beans, wheat, and cotton are the most productive, whereas sugarcane, rice, short berseem, and corn seem to be the most wasteful of water.

Because of the absence of certainty as to whether it is water or land that is the ultimate determinant of scarcity, the third column in Table 12.2 combines the two sets of indices by showing their geometric average. Again we find that because productivity dispersion is more pronounced

Table 12.1   Indices of Crop Productivity of Land and Water in Egyptian Agriculture, 1990/91

Crops	Value Added/Land	Value Added/Water
Perennial Crops		
Sugarcane	0.75	0.33
Orchards	0.70	1.17
All Season Crops		
Vegetables	1.75	1.40
Tomatoes	2.33	2.33
Main Winter Crops		
Wheat	1.00	1.89
Long Berseem	0.77	1.25
Short Berseem	0.50	0.67
Beans	1.00	2.00
Main Summer Crops		
Corn	0.93	0.93
Rice	1.25	0.38
Cotton	1.27	1.56
Potatoes	2.00	1.00
Other Crops	0.75	1.50
All Crops	1.00	1.00

Source: Based on data collected from Egyptian published sources.
Note: Numbers in the table are produced by dividing the percentage contributed by each crop to agricultural value added by the crop's percentage share in land- and water-use, respectively.

with water than with land, water obviously dominates. Tomatoes, vegetables, beans, and potatoes, followed by cotton, top the list, and sugarcane, short berseem, rice, and corn appear at the bottom.

This tentative, approximate calculation, with all its weaknesses, indicates that we should seriously consider the elimination of sugarcane and perhaps reduce rice and corn and certainly seek to expand the production of the high-value crops that are economical in the use of land and water—those at the head of Table 12.2. This is not a startling result, and certainly it is not original, but the data seem to support this view. I emphasize that the data used may not be very accurate. There is no harm in taking them at their face value at this stage, later seeking better measurements and more profound analysis by Egypt's agronomists and economists.

A Garden, Not "Security"

The obvious implication of the foregoing analysis is that Egypt should use its scarce resources more efficiently, not to become self-sufficient in food

Table 12.2 Ranking Egyptian Crops in Ascending Order of Their Land and Water Requirements (Per Unit of Value Added)

Land	Water	Land and Water Combined
Tomatoes	Tomatoes	Tomatoes
Potatoes	Beans	Vegetables
Vegetables	Wheat	Beans
Cotton	Cotton	Potatoes
Rice	Other crops	Cotton
Beans	Vegetables	Wheat
Wheat	L. Berseem	Orchards
Maize	Orchards	Other crops
L. Berseem	Potatoes	L. Berseem
Other crops	Maize	Maize
Sugarcane	Sh. Berseem	Rice
Orchards	Rice	Sh. Berseem
Sh. Berseem	Sugarcane	Sugarcane

Note: The ordering is based on data in Table 12.1. Ranking in the third column is derived from the geometric averages of the separate indices for land and water as in Table 12.1.

but to maximize economic returns to water and land. The goal of *food security*, so often stressed as a primary objective of agricultural policy, needs to be critically examined. Food is not a determinate and unchangeable collection of items. It is variable, in reaction to changes in relative prices and incomes. In times of hardship we cannot rigidly keep our pattern of consumption static, if for no other reason than that our incomes will not permit it. That is why substitution between elements of food should be contemplated when food security is addressed. Less emphasis on meat, with its excessive demands on water and land, is warranted, and stress should be placed instead on fish, pulses, and legumes, which are good for the soil, high in nutrition, and parsimonious in water-use. Corn, wheat, rice, millet, and other grains can all be thought of as substitutes, providing security if need be. We should think of food security not in terms of procuring specific foods, whatever they are, but through building stocks of financial reserves, to be held in foreign exchange that can be used to purchase variable products in the cheapest international markets. Thus should the concept of food security be viewed, interpreted, and put into practice.

Once agriculture has rid itself of such policy constraints as food security, and moved freely to optimize the use of the country's scarce resources, the possibilities are immense to turn Egypt into a huge productive garden using modern techniques and organization methods, producing fruits and vegetables, flowers, ornamental plants, and other highly valued crops, taking advantage of the availability of labor that is likely to become more and more abundant as its numbers and productivity rise. The value added created by this new pattern of production, and by the ancillary services that must be developed to ensure a successful export drive of the new products, will be immense.

The technology, industrial and commercial organization, and investments necessary for turning the agricultural sector into this highly productive machine should not be underestimated, but they are certainly within the capacity of the Egyptians. Exports will then be the engine of sustaining this pattern of production, which will require vast complementary activities for sorting, standardizing, and packaging the products, and marketing them at home and abroad. Investment in such activities, including market research, should be accorded the highest priority and promises to yield considerable returns and generate externalities that would carry through to the markets of manufactures.

Markets abroad need to be located and developed through intelligent market research, and sustained through business and diplomatic contacts and continuous uplifting of quality standards. For this the help of the government is essential, not just to ensure adequate infrastructure, the availability of finance, and various encouragements of exports, but principally to negotiate the entry of Egypt's agricultural produce, whether raw, canned, or frozen, to markets that are usually unreceptive. In this regard fishery products should not be forgotten, especially the products of aquaculture, and exploiting the immense fisheries wealth that can be tapped from the Red Sea and from the Nile on a sustainable basis.

The Price of Water: A Suggestion

Finally, a further word on the issue of water pricing: with the new market orientation of the Egyptian economy, and with freeing the prices of agricultural inputs and outputs, water must carry a price commensurate with its utility and scarcity, since this will direct it to its more productive uses. Difficulties, it is true, stand in the way of charging a price for water-use in agriculture. There is first the emotional reaction that such a vital commodity should not be marketed—water being a "gift of God" as the argument goes. Second, there is the difficulty and expense of metering water—a problem made doubly difficult by the fragmentation of landholdings. Third, the topography of agricultural land in Egypt creates hydraulic interdependence among farms in the sense that lower lands may receive for irrigation drained water already used by higher placed farmers.

In the face of this resistance to the pricing of water used in agriculture, a way out might be to go back to the old system of imposing a levy on the final price of each crop. What is needed here is a set of water charges to raise enough resources to cover at least the variable costs of providing irrigation water and draining it successfully. These charges could relate to the theoretical water requirements of each crop, and be collected with the land tax after the annual cropping pattern has been established. An acre under beans requires a fraction of the irrigation water required by an acre of sugarcane, and the tax levied on beans and sugarcane would reflect this fact. In this way, water would be charged indirectly. Work should proceed, however, to explore the possibility of charging directly for water in agriculture. This would be the more efficient way of tackling the problem.

References

Hansen, Bent. *The Political Economy of Poverty, Equity, and Growth: Egypt and Turkey.* New York: Oxford University Press, 1991.

World Bank President's Report (No. P-5560-EGT), "A Proposed Structural Adjustment Loan to the Arab Republic of Egypt," November 1991.

World Commission on Environment and Development. *Our Common Future.* New York: Oxford University Press, 1987.

13

Strategy of the Commons: Opportunities and Challenges for Egyptian Agriculture

Joseph R. Potvin

About 70 percent of the mass of every living organism is water, and in Egypt most of that water comes from the Nile. Dr. Boutros-Ghali, Secretary-General of the United Nations, once commented that "The national security of Egypt, which is based on the waters of the Nile, is in the hands of eight other African countries"—the countries from which the river flows (Roberts 1991: 20). Most of these countries are anxious to finance construction of hydroelectric dams, with foreign capital in the hands of some twenty-five potential lending countries and institutions. The future of Egypt is also entwined in the greenhouse politics of 165 other countries. For Egypt is one of the nations most seriously threatened by the prospect of a rise in sea level (the Mediterranean included) from global warming induced by humanity's influence on the atmosphere. The Intergovernmental Panel on Climate Change (IPCC) expects a global mean rise in sea levels of 6 cm per decade over the next century (Houghton et al. 1990 xxix). Perhaps a fifth of Egypt's most densely populated and productive agricultural areas could be flooded within a hundred years.

Egyptians obtain a portion of their drinking water from the Nubian groundwater aquifers that drift under Egypt and three neighboring countries. In each of the countries with direct access to the aquifers, there are also groundwater well systems for irrigation and several proposals toward construction of further installations (Attia and Lennaerts 1989). But unconstrained extractions could drastically lower groundwater levels and render this resource brackish and uneconomic in fifty years. They would take thirty thousand years to recharge (Roberts 1991: 21). Fortunately, none of these three water problems can be solved by war. The only route that offers realistic prospects for management is multilateral collaboration for the conservation of resources and the environment. Furthermore, several existing international accords, institutions, and trends together furnish practical mechanisms that can help to secure the necessary collaboration and foster compliance.

155

The second section of this chapter considers three routes by which Egypt's vital interests in the conservation of shared environmental infrastructure can be advanced. One of the potential elements of a workable solution, if member states would agree to harmonize resource efficiency and environmental standards, is the General Agreement on Tariffs and Trade (GATT). A second avenue to sustainable water management is the integration of ecological conservation and resource efficiency conditions in determinations by international financial institutions of the creditworthiness of prospective borrowers. A third opportunity is presented in greater environmental awareness and activism among consumers, workers, and business leaders around the world, at least some of whom would volunteer to forego imports or exports associated with the destruction of common resources or the degradation of ecosystem integrity. Each of these developments provides both vital opportunities and tough challenges for Egypt's agricultural sector towards the next century.

The third section of the chapter briefly considers Egyptian agriculture with regard to ecological impacts. Some very general suggestions are made for agricultural reforms to strengthening its international bargaining position for the conservation of shared freshwater supplies, and the reduction of risks and damages from global climate change.

Trends in International Commerce

In February 1992, Arthur Dunkel, GATT director general, announced the release of a discussion paper, *Trade and the Environment* (GATT 1992), calling it the organization's "first authoritative attempt" to address issues and concerns about potential formalization of resource efficiency and environmental standards in the administration of international trade relations. A few months earlier, the Group on Environmental Measures and International Trade in the GATT also reconvened after two decades of inactivity to assess prospects and formulate options. Existing regulations under the GATT allow member countries the flexibility to apply differential treatment to goods and invoke trade-restricting measures if this is necessary "to protect human, animal or plant life or health" (Article XX, b) or, for reasons "relating to the conservation of exhaustible natural resources" (Article XX, g). The GATT further condones "technical regulations or standards" for the protection of "human health or safety, animal or plant life or health, the environment, fundamental climatic or geographical factors" (Article 2:2 of the GATT Standards Code), and allows subsidies for the "redeployment of industry in order to avoid congestion and environmental problems" (Article 11:1, f of the GATT Subsidy Code).

GATT is supposed to be a practical accord among governments

sensitive to principles of commerce by which they may promote the interests of their own people, and the interests of the international community as a whole. With regard to ecological conservation and resource efficiency, some have tried to argue that "harmonization would imply the introduction of identical standards . . . regardless of the assimilative capacity of nature and of social preferences" (Rauscher 1992: 183). Others have portrayed the move to apply conservation and efficiency standards as disguised protectionism by established producers. Whenever any class of trade restrictions is wielded *unilaterally*, there is risk of legitimate issues being "kidnapped by trade protectionists" (Dunkel 1992).

The two central concerns currently being expressed by the GATT officials are that conservation and efficiency standards applied to trade should (1) be common and general, not unilateral and ad hoc, and (2) not discriminate between the home-produced goods and imports, nor between imports from and exports to different trading partners (Dunkel 1992). To come up with an accord that solves these two dilemmas constitutes the strategic design problem upon which the effort to link trade and environment is apt to succeed or fail. "The real challenge we face," observes Dunkel, "is to forge a constructive alliance between trade liberalization and strengthening the multilateral trading system on the one hand, and environmental protection and conservation on the other."

A variety of proposals have recently been raised by member governments advocating multilateral negotiation of more comprehensive rules. They would allow member countries with more rigorous conservation and efficiency standards to impose special tariffs to offset lower production costs of goods from countries with less stringent standards (for example, see *Inside U.S. Trade* 20.9.91). Haavelmo and Hansen (1991) have argued that "Tariffs to protect an efficient national policy of cost internalization (not an inefficient industry) should not be ruled out as unwarranted 'protectionism.' Unpaid environmental costs . . . are subsidies reducing the price of exports—tantamount to dumping." Within ten to fifteen years, the Polluter Pays Principle (PPP)—or perhaps more generally a "Perpetrator Pays Principle"—will be formally and forcefully incorporated in the operational precepts and procedures of most major multilateral economic and legal entities. Included among them will be the GATT.

International Finance

Creditworthiness has never been based on simple calculations. In addition to quantitative financial appraisal, it has always involved a set of qualitative reckonings about the human and infrastructural context of the borrower. At the smallest scale, a variety of microenterprise credit institutions around the world have demonstrated the strength of sensitive human judgments in sound financial management, when borrowers lack the barest physical or financial collateral (Christen 1990). Now, after the manifest failure of so many large-scale econometric formula-driven debt financing strategies, and with the benefit of post-Brundtland perspectives, the lending criteria of some of the major multilateral and commercial banks are attaching more weight to borrowers' "overall creditworthiness." Matthews and Contreras-Manfredi (1989: 4) propose that the cumulative assessment of human, natural resource, biophysical, and financial capabilities provides a more reliable indicator of a prospective client's potential for honoring financial obligations. Development finance institutions and commercial banks are both taking steps to ensure that the investments they help to finance are ecologically sound.

It is now the norm that international development finance banks and aid agencies require recipients of project support to adopt and comply with a set of prespecified land-use conditions and practices. The World Bank, for example, recently published a three-volume *Environmental Assessment Sourcebook* with specific sectoral guidelines (World Bank 1991a/II) and detailed recommendations for energy and industry projects (1991a/III). Since 1973 Egypt has consistently carried a serious balance of payments deficit, with an annual commodity import burden of up to two and three times the value of its commodity export revenues. More than 20 percent of Egypt's import expenditures go to supply half of the country's food. While considerable progress has been made over the last several years to boost annual export revenues faster than the import bill, imports still exceed exports by US$7 billion annually. After private transfers to the country of $4 billion, there remains a US$3 billion current account deficit, US$2 billion of which is made up through foreign and multilateral development assistance (World Bank 1991b: 231).

The private sector financial institutions are also changing as environmental standards take on increasing importance in appraisals of client creditworthiness. This has been variously referred to as the Valdez Effect, or the Bhopal Effect. This can seriously affect cash flow for individual firms, as well as for entire industrial sectors and countries (Conklin et al. 1991: 112). In this event, the borrower's ability to service debt is jeopardized, and sources of new credit disappear. The bulk of Egypt's foreign commercial financing is with German and French banks, which are among the leaders in developing procedures that reflect this new code of accountability.

International Market Characteristics

In addition to weighing current trends in the GATT and among international financial institutions, Egyptian producers may wish to consider how best to position themselves: (1) in an export market that is quickly moving to distinguish commodities with regard to the ecological impacts and risks of their production processes; and (2) as importers from markets that are in the midst of structural reforms to implement "full-cost pricing" according to the PPP. About 40 percent of Egypt's export and import life today is coupled to the European Community. In 1988 Egypt's two single largest export destinations were the USSR, with 12.2 percent, and Italy, with 11.1 percent of the total. The two largest sources of its imports were the United States, supplying 11.9 percent, and Germany, with 11.1 percent. A decade earlier, Italy accounted for over 27 percent of the exports, and the United States supplied almost 18 percent of the imports (United Nations 1991: 260).

The particular way in which resource efficiency and ecological concerns arise in the European market for Egyptian products must inevitably manifest itself among the strategic economic concerns of Egyptians. Organically produced foods command premium prices (often 100 percent higher than conventionally produced foods) and they are more labor intensive. These financial and employment advantages should not be overlooked. There is little doubt that among Egypt's most important trading partners, consumers, workers, and corporate managers are choosing to weigh the ecological soundness of the things they buy and produce, not absolutely but to a significant degree.

Challenges Facing Egyptian Agriculture

It follows from the preceding section that environmental and resource efficiency standards applied through multilateral trade rules, lending criteria of international financial institutions, and trends in global market demand could provide effective support for the conservation of Egypt's shared freshwater supplies and help to avert some of the calamity expected with continued increases in global "greenhouse gas" emissions. Yet collaboration in the conservation of shared environmental systems is truly a package deal, involving many intersectoral and transectoral issues. For instance, the strong shift of Egypt's electricity supply in the last fifteen years, toward generation systems powered by fossil fuels, counters the country's own strategic interests in having all countries act quickly to minimize the causal influences of sea-level rise. Egypt's own annual anthropogenic additions to three principal "greenhouse gases" (CO_2, methane, CFC) in the latter 1980s was 17 million equivalent tonnes of CO_2. The conservation ethic affords vital opportunities, but also raises tough challenges for Egypt's economy and society. As wide-ranging conservation standards come to be implemented more generally and with more diverse enforcement channels, it is important to contemplate how various facets of the country's life are influenced and how people might respond. Among the most ubiquitous concerns are resource throughput and population growth. Throughput is meant in terms of the relative mass-energy dimensions emphasized by Daly (1991). Nations of the North are notoriously inefficient consumers, using and wasting many times the resources per capita as do the citizens of the South. Conservation strategies in the South should involve radical reductions in the rates of population expansion, while taking advantage of every opportunity to improve resource-use efficiency, and reduce waste. Population growth can be expected to play a role in global standards. With a current population of 55 million and a 2.6 percent annual growth rate (3.3 percent in urban and 1.8 percent in rural areas), Egypt is expected to have 70 million people ten years from now (Bulatao et al. 1990).

To illustrate some challenges facing Egyptian agriculture in particular, the following is a consideration of some of the available indicators of Egypt's relative international ranking in the intensity of use of chemical inputs on cropland.

Chemical Inputs to Cropland

Fertilizers: The nitrogen and phosphorous runoffs from concentrated fertilizers from farmland into river systems produce aberrant surges in algae, plant, and microbial growth, which leads to severe local or regional oxygen depletion of the water. "Eutrophication," and its consequent mass kills of fish, is frequent throughout the world, and getting worse (WRI 1990: 182). Furthermore, chemical fertilizers used in agriculture (especially in conjunction with chemical pesticides) fail to maintain the biological richness of soil quality that is obtained with the organic compost matter and manures that they are presumed to replace.

Egyptian farmers have almost doubled the intensity of their fertilizer application between the mid-1970s and the mid-1980s. Now at 347 kilos per hectare per year, this is 30 percent more than the European average. Although it remains comparable with some individual European countries, Egypt nevertheless has the most fertilizer-dependent agricultural system of all the developing nations (WRI 1990: 280–281). This status becomes a competitive disadvantage as key markets and international sources of finance capital grow intolerant of the associated ecological damages, and as important trading partners begin to change their own production systems. Among trade analysts, it is now considered that

> When farmers find it cheaper to forego sustainable soil conservation farming practices by adding subsidized chemical fertilizers to offset loss of topsoil and depletion of desirable soil properties, the true cost of non-sustainable cultivation practices is not taken into account in the choice of farming technology and input mixes (Hansen 1988: 16).

Pesticides: Organochlorine insecticides and fungicides are among the most persistent and widespread toxic contaminants of water systems (Environment Canada et al. 1991: 4). By impairing the normal biochemical and physiological functions of living things, their individual and combined effects are responsible for increased deformities, cancerous tumors, and reproductive abnormalities, as well as poorly understood behavioral changes. During the 1970s and 1980s, reliance on poisons in Egyptian agriculture was much greater than it would have been if left to conventional market forces: in the early 1980s the government was spending US$200 million per year to subsidize 83 percent of their full retail cost (Repetto 1985). This was one of the world's highest rates of subsidy for pesticides: in 1982 the Egyptian government spent more per capita on pesticide subsidies than it did on health care (Postel and Flavin 1991: 178). Nevertheless, over the last ten years pesticide intensity has been reduced by 25 percent, now averaging 7.6 kilos of active ingredient per hectare of cropland. This is about the same as in Greece, although the average application intensity across Europe is still much lower at about 4.8 kilos per hectare (WRI 1990: 280–281). Improved standards and controls over the use of pesticides for agricultural production, disposal of pesticide wastes, licensing and training of pesticide users, and closer monitoring of river and groundwater quality must be considered important factors in the positioning of Egypt's agricultural trade, finance, and marketing strategies for the 1990s.

Strategic Positioning

In the broadest sense, export revenues can be raised by increasing the volume exported, or by increasing the quality, or both. Recognition of the biophysical impossibility of forever increasing production and consumption volumes while maintaining the integrity of ecosystem infrastructure suggests that sustainable development depends on quality rather than volume. Egypt should not put its energies into boosting agricultural exports by boosting volume. Instead, a medium- and long-term strategy should be to boost the *quality* of the export commodities. Since many of Egypt's principal export markets are now distinguishing goods according to the ecological soundness of their production methods, changes in processes to reduce pollution make the goods more attractive, and constitute an actual improvement in their quality. It may be useful to review agricultural assistance programs to ensure that they are not actively discouraging ecologically favorable practices, such as when farm income support programs are tied to output volume (OECD 1989: 68). Regarding production for domestic demand, the predominance in Egypt of traditional smallholder family farms that produce both for sale and subsistence, often combining crops and animals, provides an excellent foundation for conversion to sustainable farming (MacLaren 1980: 5–6). Experience in many countries suggests that advisory and extension services can be used both to increase farm income by reducing expenditure on commercial inputs, and decrease agricultural pollution (OECD 1989: 66).

References

Attia, F.A.R., and A.B.M. Lennaerts. "Economic Aspects of Groundwater Development for Irrigation and Drainage in the Nile Valley," in *Groundwater Economics*. Edited by E. Custodio and A. Gurgui. Amsterdam: Elsevier, 1989.

Bulatao, R.A., E. Bos, P.W. Stephens, and M.T. Vu. *World Population Projections, 1989–90 Edition*. Baltimore: Johns Hopkins University Press, 1990.

Christen, R.P. *Financial Management of Micro-Credit Programs*. Cambridge, Mass: ACCION International, 1990.

Conklin, D.W., R.C. Hodgson, and E.D. Watson. *Sustainable Development: A Manager's Handbook*. Ottawa: National Round Table on the Environment and the Economy, 1991.

Daly, H.E. "Towards an Environmental Macroeconomics," *Land Economics*, 1991, 67: 255–259.

Dunkel, A. "Asia and the Pacific: Merging Business and the Environment." Address to the Conference of the International Herald Tribune in Association with the Thailand Development Research Institute. Bangkok, 1992.

Environment Canada, Department of Fisheries and Oceans, and Health and Welfare Canada. *Toxic Chemicals in the Great Lakes and Associated Effects*. Ottawa: Supply and Services Canada. Cat. No. En37-94/1990E. 1991.

GATT. *Trade and the Environment*. Geneva: Secretariat of the General Agreement on Tariffs and Trade, 1992.

Haavelmo, T., and S. Hansen. "On the Strategy of Trying to Reduce Inequality by Expanding the Scale of Human Activity," in *Environmentally Sustainable Economic Development: Building on Brundtland*. Edited by R. Goodland, H. Daly, and S. El-Serafy. Washington D.C.: World Bank, Environment Department Working Paper No. 46. 1991.

Hansen, S. *Structural Adjustment Programs and Sustainable Development*. Paper commissioned by the United Nations Environment Program (UNEP) for the annual session of the Committee of International Development Institutions on the Environment (CIDIE), Washington D.C.: The World Bank, June 1988.

Houghton, J.T., G.J. Jenkins, and J.J. Ephraums. *Climate Change: The IPCC Scientific Assessment*. Cambridge: Cambridge University Press, 1990.

MacLaren, J.F. et al. *Lake Manzala Study*. Vol. 1. Willowdale, Canada: Ministry of Development and New Communities, Egypt, and United Nations Development Program, 1980.

Matthews, G., and H. Contreras-Manfredi. "Sustainable Development, Creditworthiness, and Quality of Life." Washington, D.C., 1989. Mimeographed.

OECD. *Agricultural and Environmental Policies: Opportunities for Integration*. Paris: OECD, 1989.

Petersmann, E. "Trade Policy, Environmental Policy and the GATT," *Aussenwirtschaft*, 1991, 46: 197–221.

Postel, S., and C. Flavin. "Reshaping the Global Economy," in *State of the World*

1991. Edited by L.R. Brown et al. New York: W.W. Norton, 1991.

Rauscher, M. "International Economic Integration and the Environment: The Case of Europe," in *The Greening of World Trade Issues*. Edited by K. Anderson and R. Blackhurst. London: Harvester Wheatsheaf, 1992.

Repetto, R. *Paying the Price: Pesticide Subsidies in Developing Countries*. Washington, D.C.: World Resources Institute, 1985.

Roberts, H. "Deep Waters Run Still," *South:* No. 124, August 1991, 19–20.

United Nations Statistical Office. *1989 International Trade Statistics Yearbook*. Vol I. New York: United Nations, Department of International Economic and Social Affairs, 1991.

Walter, I. *Environmental Resource Costs and the Patterns of North-South Trade*. Paper prepared for WCED, 1986.

World Bank. *Environmental Assessment Sourcebook*. Vol I: Policies, Procedures, and CrossSectoral Issues; Vol II: Sectoral Guidelines; Vol III: Guidelines for Environmental Assessment of Energy and Industry Projects. Washington, D.C.: World Bank, Environment Department, 1991a.

———. *World Tables 1991*. Baltimore: Johns Hopkins University Press, 1991b.

World Resources Institute. *World Resources*. New York: Oxford University Press, 1990.

14

The Uruguay Round of Multilateral Negotiations and Egyptian Agriculture

Mohamed Maamoun Abdel Fattah

The Uruguay Round of multilateral trade negotiations is the eighth round of negotiations since the GATT (General Agreement on Tariffs and Trade) came into being in 1947. GATT is not part of the United Nations system; it is, however, the organization in charge of the world trading system. It has now 108 member countries covering over 95 percent of the world trade. Members in GATT are called the "contracting parties," because of the contractual nature of the organization, and members have to secure their membership through (1) submitting a schedule of tariff reductions and (2) abiding by the GATT rules of free trade in their trade policy. It took about eight years of negotiations with GATT for Egypt to gain its membership in 1970.

Although GATT articles deal with all types of merchandise trade, agriculture has not received the full extent of trade liberalization as have the other sectors of international trade. Despite the efforts of GATT on problems of trade in agriculture in the previous rounds, it failed to achieve positive results. Agriculture remained without discipline and rules, unlike the industrial sector, because of the complexities of agriculture in domestic policies and international trade. Agricultural trade was still under higher tariffs and encountered all types of nontariff barriers. As a result, agriculture faced a variety of trade distortions and frictions started to spread, affecting many developed and developing countries in varying degrees.

Developed countries have increased their agricultural producer subsidies progressively in the last three decades as shown in these OECD figures, cited in *The Economist*, December 12, 1992:

Region/Country	Average Annual in US$ Billions	
	1979–1986	1991
EEC	38	83
Japan	22	30
United States	30	34

165

The overall producer subsidies in the developed countries amounted to US$170 billion in 1991. The farm subsidy was equivalent to 30 percent of the farmers' income in the United States, 49 percent in the EEC, and 66 percent in Japan. Worldwide, the costs of farm protection are even greater. Several studies (cited in *The Economist*, September 22, 1992) have shown that a complete liberalization of trade in agricultural goods would expand trade by US$100 billion per year, of which the potential increase in the agricultural exports from developing countries would be nearly US$30 billion. The Uruguay Round of GATT negotiations has witnessed a world-wide consensus that agricultural trade has to be reformed and subjected to an equal discipline as with other sectors involved in international trade.

The negotiating objective in agricultural trade was that the contracting parties agree that there is an urgent need to bring more discipline and predictability by correcting and preventing restrictions and distortions, including those related to structural surpluses, so as to reduce the uncertainty, imbalances, and distortions in world agricultural markets. The negotiations aim to achieve greater liberalization of trade in agriculture and bring all measures affecting import access and export competition under strengthened and more operationally effective GATT rules, taking into account the general principles governing the negotiations by:

- Improving market access through, inter alia, the reduction of import barriers
- Improving the competitive environment by increasing discipline on the use of direct and indirect subsidies and other measures affecting directly or indirectly agricultural trade, including the phased reduction of their negative effects and dealing with their causes
- Minimizing the adverse effects that sanitary and phytosanitary regulations and barriers can have on trade in agriculture, taking into account the relevant international agreements

Negotiations under the Uruguay Round have been held since the Punta del Este declaration in September 1986. A draft agreement was submitted to the negotiators for acceptance in December 1991. Some details in this agreement have yet to be finalized. It must be mentioned— and it is quite well known—that the political decision to accept the text on agriculture (and the rest of the agreement) in the Final Act of the Uruguay Round is still pending.

Egypt: Agriculture's Changing Role

Agriculture in Egypt plays an important role in the economy. At present its share in the gross domestic product (GDP) is about 20 percent; it engages about 37 percent of the labor force; and contributes about 20 percent of export earnings as well as important inputs to the manufacturing sector. The importance of agriculture to the Egyptian economy is likely to increase in the future because of the positive effects of the policies adopted by the government since 1986 to restructure the sector, in cooperation with international financial institutions and within the framework of a comprehensive structural reform and economic liberalization program. The new agricultural policy based on decentralization and deregulation has to be viewed in the light of overall macroeconomic policies, where it affects and is affected by the economic reform policies such as liberalization of trade, exchange rate, interest rate, reducing the fiscal deficit, and privatization.

Agricultural policies in Egypt in the past were biased in favor of the industrial sector and the consumer, focusing on the social objectives of providing the people with foodstuffs at low prices. These policies led to dramatic changes in production, investment, and consumer behavior. Until 1970 Egypt grew enough food to feed its population. This picture has changed in the last twenty years and Egypt has become one of the major food importing countries. The agricultural import bill soared to US$3.7 billion in 1990, and became quite vulnerable to international prices. The imports covered wheat, wheat flour, corn, meat, poultry, fish, dairy products, sugar, edible oils, and even beans and lentils. It can be argued that Egyptian agriculture cannot sustain feeding its fast growing population while agricultural resources are limited: cultivable land has reached its limitations; water resources are scarce. The fact remains that the bias against producers had diminished the capacity to efficiently utilize these scarce resources.

The bias against agriculture was reflected clearly in the following measures:

- Very low protection was given to domestic agricultural producers: some food products—e.g., wheat and corn—are still imported at no more than 1 percent tariff. The average tariff on agricultural imports is estimated to be around 19 percent, while the industrial products received tariff protection of over 100 percent.
- The exchange rate—on which the value of imports are calculated—was almost one-third of its actual value. This anomaly was corrected only in October 1991.
- Consumer subsidies for food had reached very high levels, which were simply subsidizing imports or conferring the benefits of subsidies to the foreign farmer.

While the farmers in developed countries were heavily subsidized, Egypt and many other developing countries were taxing their farmers (Dethier 1991). According to a World Bank study (1992), the implicit tax on crops in Egypt was estimated at £E 5.5 billion in 1985 but had fallen to £E 1 billion in 1991. Similarly, in studies undertaken by Egypt's Ministry of Agriculture, with assistance from the GATT, it was found that the producer subsidy equivalents were negative for all major crops (Dethier 1991).

Beside the foregoing implicit tax, agriculture in Egypt remained under heavy government regulation and even direct intervention. This was manifested by a whole range of policies and measures, notably pricing policies which relied more on administrative decisions than market forces; compulsory membership in cooperative marketing arrangements intro-duced in 1962; and mandatory delivery quotas that were established to ensure supply of agricultural products at low administrated prices. It is fair to add that the government supplied the farmers with inputs, such as fertilizers, pesticides, and selected seeds, at subsidized prices, but this did not offset the impact of the negative measures on farmers' incomes and incentives.

The policymakers in Egypt became fully aware of the distortions and the disequilibria caused by these policies. Substantial progress has been made in adjusting the agricultural sector policies: pricing of crops has been improved and the crop area allotments and delivery quotas have been eliminated except for cotton, sugarcane, and part of the rice crop. These policies will be completed in the near future. Thus agriculture will take the lead in being deregulated and fully privatized before the other sectors of the economy. It should, however, be noted that Egyptian agricultural policies cannot be formulated in isolation. They have been affected by the external environment and will continue to be affected by policies and measures adopted by other countries.

Egypt's change from being a self-sufficient country before 1970 to being one of the largest importers of foodstuffs in the world, is not necessarily because of its domestic policies alone but also because of the availability of food products at subsidized prices. Developed countries have acknowledged that their policies have harmed agriculture in many developing countries. True, Egyptian consumers have enjoyed cheap food, but Egyptian farmers and society in general have paid a high cost. On the other hand, Egyptian agricultural exports have been affected by the multitude of barriers, both tariff and nontariff. The potential is great for Egypt to export fruits and vegetables and other agricultural products if these barriers are dismantled.

The Negotiations

The long-term objective set out in the draft GATT agreement is to establish a fair and market-oriented agricultural trading system. The reform process would be initiated through commitments on support and protection and the establishment of strengthened and more operationally effective GATT rules and disciplines. These objectives are to be implemented through specific commitments in four areas: (1) market access; (2) domestic support; (3) export competition; and (4) agreement on sanitary and phytosanitary issues. Due regard is to be given to various groups of countries when they assume commitments for:

- Nontrade concerns and food security
- Special and differential treatment to developing countries
- Taking into account the possible negative affects of the reform program on net food-importing developing countries

The product coverage of the draft agreement has been settled—after prolonged discussions—to include chapters 1 to 24 of the Harmonized System of Customs Nomenclature or Classification, plus certain other agricultural products, like wool, hides and skins, cotton, and flax, that fall under other tariff headings. The coverage excludes fish and fish products as they will be treated in the domain of other market-access commitments.

Market Access

Countries accepting the agreement are committed to the following:

1. Reduce their bound customs duty by 36 percent over a period of seven years from 1993 to 1999. The 36 percent reduction would be on the level applied in September 1986. (The bound tariff duty in GATT language means that duty cannot be raised more than what countries submitted in their schedule of concessions.)
2. For products which are subject to other measures at the border—nontariff barriers—these barriers will be transformed or converted into customs duties. These too will be reduced by 36 percent over seven years. This transformation is called "tariffication" and is calculated by modalities which can be summarized as the difference between the (CIF) import prices and the average wholesale price in the domestic market, based on the period 1986–1988.
3. The developing countries would receive special and differential treatment: they are allowed to assume commitment over a period of ten years and the rate of reduction would be 24 percent.

The tariffication and reduction commitments as described may cause injury to agricultural production. A special safeguard mechanism is stipulated in the text to allow countries to use it if needed.

Domestic Support

Domestic support is a form of subsidy given to producers. Domestic subsidies are not forbidden in the GATT rules and they have been used widely in a manner injurious to international trade. It is claimed that subsidies increase production and surpluses in a manner that increases exports, or inhibits imports, or both. The vagueness of the present GATT rules has led to the new text, which tries to clarify in detail what is allowed and what is forbidden in subsidy policy.

Because of the variety of domestic supports—in form and method of payment and the different names given to them in each country—the text has created a measurement to quantify them. It is called the Aggregate Measurement of Support (AMS).

Countries accepting the agreement are committed to reduce domestic support by 20 percent over seven years from 1993 to 1999. Developing countries are required to reduce two-thirds of that 20 percent. If it is found that the domestic subsidy does not exceed 5 percent of the total value of production, it will not be subject to reduction commitments (10 percent in developing countries).

Not all forms of domestic support are subject to reductions. The agreement has a list of subsidies which are exempt from the reduction commitments, including government general services in the areas of research, pest and disease control, training, transfer of information to producers and consumers, infrastructure, and inspection, marketing, and promotion. They also include payments by governments to improve the income of producers, but it should not be linked to type or volume of production (decoupled income support); insurance and relief from natural disasters; structural adjustment assistance; and environmental programs.

Export Competition

Export subsidies are forbidden under the GATT. However, a number of conflicts and disputes have arisen in past years because "certain primary products" were allowed for. This exemption led to all types of export subsidies in agriculture.

Countries accepting the agreement are committed to reduce their export subsidies by 36 percent of the 1986–1988 level over a period of seven years from 1993. Developing countries are allowed ten years and 24 percent (i.e., two-thirds of the 36 percent).

The agreement lists the export subsidies subject to reduction commitments and defines each measure in a manner designed not to be subject to misinterpretation in the future.

Sanitary Measures

The draft agreement on agriculture incorporates a separate agreement on sanitary and phytosanitary measures. It is designed to ensure the right of every country to draft and apply rules for the protection of human, animal, and plant life. However, rules also inhibit misuse to disguise intentions of unjustifiable discrimination between countries for restricting international trade.

Reform and Egypt's Agricultural Trade

Egypt has long been an active party to the international consensus in GATT to start the process of reform in agricultural trade. The reform would serve its interests in the long run, but in the short run Egypt's trade bill will be affected adversely through increased prices resulting from the elimination of subsidies, especially in the major producing countries like the United States and members of the EEC. Research institutes and international organizations agree that prices will go up, but estimates differ about how much. A possible scenario is that the rise will be almost 30 percent. This could mean an additional US$1 billion in Egypt's trade bill for the import of foodstuffs—although the new projected reductions in support and protection mentioned in the draft agreement would reduce this estimate to about US$300 million. The Egyptian negotiators in Geneva convinced the GATT negotiating group that Egypt would be a net loser from the reform program in agricultural trade in the short run. During the transitional period, ways should be found to alleviate the possible negative effects of the reform program on developing countries that are net importers of food. Egypt is top of the list of this group of twenty-eight developing and least-developing countries.

Egypt was joined in these negotiations by other countries that faced the same problems—Jamaica, Mexico, Morocco, and Peru. These five countries worked hard to convince other countries of their justified case. They asked for a package of measures that would be implemented in cooperation with international financial and development organizations. These measures should have two main objectives: alleviate the burden of increased prices on the import bill and balance of payments situation; enhance the capacity to increase agricultural production, especially food production, taking into consideration the internal weakness of the agricultural sector in most developing countries. These objectives can be achieved through:

- Provision of increased financial resources and technical assistance, bilaterally and multilaterally, to enhance agricultural production, productivity, infrastructure, and research.

- Increased food aid, taking, inter alia, a flexible approach to the usual marketing requirement, and triangular arrangements to promote production and exports of developing countries.
- Enhancing the purchasing power—this is to be done by concessional sales, including increased availability of low-cost export credits and grants, particularly to those countries with debt servicing problems.
- Increasing export earnings by way of improved market access conditions for agricultural exports; this is to be done by immediate reduction in tariffs and phasing out or elimination of nontariff measures. Elimination of trade-distorting support and protection measures should cover products of export interest to these countries.

The case was not an easy one to make and the negotiations continued over five years. Opposition to the demands continued, and these counter-arguments were presented:

- It is not certain that prices *will* go up because of the reform program: world production will adjust to the new market situation and prices may go down and stabilize.
- The competence of GATT is restricted to trade matters and does not go into the domain of other international organizations.
- There is no generally agreed definition of a "net food-importing country." Besides, a country may change and become self-sufficient, or an exporter. Moreover, rich countries like Saudi Arabia can qualify in this group, although they are able to pay increased prices.
- Specifically which countries would pay the cost of this package?
- It is not clear how a mechanism can be created to put into effect a package, especially where it involves a number of international organizations.

The major difficulty came from countries that feared food aid could circumvent the disciplines on export subsidies. Finally, a compromise text—in the form of a declaration—was included in the draft agreement on agriculture. The language of the text provides for establishing mechanisms to ensure that the result of the Uruguay Round does not adversely affect the availability of necessary food aid to the net food-importing countries and the least-developing countries. This would be achieved through the Food Aid Convention, either as full grant or at appropriate concessional terms. The text includes provisions to give technical and financial assistance to improve agricultural productivity and infrastructure. As regards the financial difficulties, the GATT has referred the matter to the international financial institutions, to be dealt with under either existing or new facilities. Negotiations between the GATT, the IMF, and the World Bank are continuing to see how best to coordinate this matter. It is also envisaged that these countries will be given market access opportunities to increase their export earnings and so help them to pay for imports.

The draft stipulates that all participants should assume obligations, but gives developing countries a little flexibility. Egypt would assume obligations in the reduction of tariffs and the process of tariffication, and in the areas of domestic support and export competition. This is being studied by the Egyptian authorities. Taking into account the present policies of liberalizing and restructuring the economy, and given the fact that support and protection for agriculture in Egypt is not high, it may safely be said that the obligations required can easily be met during the next ten years.

The Transition

It is clear that Egyptian agriculture, which is undergoing dramatic positive changes, will be affected by the new Uruguay Round agreement on agriculture. Cheap food will no longer be available. Consequently, the interrelationship between production, consumption, and trade will take on a different perspective. Egyptian agriculture will have to respond to the new international market situation. Production will have to react to a situation involving greater dependence on the domestic market and less on imports.

The transition period for full adjustment will last about seven to ten years. Egyptian policymakers have to immediately project and plan ahead to prepare for increasing exports and to make use of the liberalization in foreign markets, not only to increase volumes (and value) but also export new products. It should also be noted that the new discipline on subsidies in *other* countries will help to make Egyptian agricultural exports more competitive.

References

Dethier, Jean-Jacques. "Agricultural Pricing in Egypt," in *The Political Economy of Agricultural Pricing Policy*. Edited by Anne O. Krueger, Maurice Schiff, and Alberto Valdés. Vol. 3. Baltimore: Johns Hopkins University Press, 1991.
World Bank. "Arab Republic of Egypt: An Agricultural Strategy for the 1990s." A Draft Report. Washington, D.C.: July 1992.

Part 6

Cultural, Social, and Political Aspects

15

The Social Sustainability of Induced Development: Culture and Organization

Michael M. Cernea

The concern for the sustainability of induced development programs has considerably increased over the last decade. However, the intellectual debate about sustainability is still largely limited to only two of its basic dimensions: economic and environmental. Granted, these two dimensions are important, but definitely these two are not the only major determinants of sustainability. This is why I appreciate the orientation of the organizers of the conference. They included a full-fledged debate on the *cultural* and *social* components of sustainable development of Egyptian agriculture. Indeed, this is crucial, because even development programs which are environmentally or economically sound may definitely stumble and eventually crumble if they are not also socially durable and enduring.

A group of my colleagues in the World Bank working on the issues of sustainability have made a somehow funny yet correct observation: they noted that sustainable development looks very different depending on whether you contemplate it from the economic or from the environmental perspective (El-Ashry 1991; Daly et al. 1992). The economic view sets much more store on accelerating economic growth, tending to explain away the possible adverse environmental consequences of economic expansion. Economists are known to argue that the negative ecological effects of growth, like pollution, social degradation, water waste, overuse of nonrenewable resources, and so forth, can be compensated due to the ability of the growth-generated surplus to pay for cleaning up the environment, or through the power of technology to produce substitutes for exhaustible materials. The environmentalists, on the other hand, emphasize the limitations inherent in nonrenewable resources, the cumulative consequences of incremental environmental damage, and urge the maintenance of the environmental services that we have inherited.

I believe that in this discussion the sociologist has to insert his or her own voice and to argue that sustainability looks yet different from his perspective: namely, sustainable development cannot be achieved through mere accumulation of technological prowess or economic expansion, and not even through prudent environmental management. Sustainability requires more than that: it demands the creation and continuous recreation of adequate *patterns of social organization* within which technological progress can unfold properly, the use of natural resources can be managed soundly, and the social actors of development, individually and collectively, can participate and share in the goals and benefits of development.

Recognizing the cultural and social prerequisites for sustainability is not a simple add-on by social scientists eager to join the debate. It is intrinsic to the very essence of the development process. Culture is humanity's basic mechanism for *adaptation* to nature, as well as for transforming it. By culture we understand, in the sociological sense, a broader set of elements than what is usually equated with culture: it is not just art and literature, or values and beliefs. Culture, as a category, is the polar opposite of nature, and is everything humankind has added to nature. Patterns of social organization are culture; in fact social organization is the *core* of human culture. It is therefore very difficult to separate the cultural from the social aspects of sustainability, but the fact that the conference had two plenary sessions for them rather than only one made it possible to discuss the issues more comprehensively.

Under the rubric "social sustainability," the aspects most frequently mentioned are participation and equity; under "cultural sustainability," reference is made to values and the need to recognize cultural diversity. In line with my emphasis on the patterns of social organization, I would like to bring up some less frequently discussed dimensions—namely, the concept of *organizational intensity* of development strategies, and the concept of *organizational density* of the social environment.

To achieve sustainability, development strategies must be not only technology-intensive, but also organization-intensive. Organization-intensive strategies lead to constructed social environments with higher degrees of organizational density. The higher the organizational density and the better the fit between organization, technology, and the requirement for managing natural resources, the higher the likelihood for sustainable development.

Induced Development in Agriculture

When we discuss the sustainability of development, we refer most often to programs for inducing development and not so much to what can be called spontaneous development. By induced development I have in mind development programs which are initiated purposively by governments, development agencies, and the like, and which use financial resources as the foremost trigger of development. For example, the many government projects financially assisted by the World Bank, or other multilateral and bilateral donors, are a case in point: they are programs in which the injection of exogenous financial resources is intended to accelerate the pace of development.

It has been observed, however, that many such government programs hardly survive the day when the inflow of exogenous financial resources ends. While aiming to launch development that is expected to continue after the closure of the original program, some programs prove unable to bring about continuous development, beyond a limited time. This is how the question of long-term sustainability first came to attention. Subsequently, this concern was reinforced by environmentalists, who legitimately pointed out that if natural resources are used up in development programs at a rate faster than these resources can be renewed, their finite character will undercut and preclude the very development they are making possible. As a result of these and other arguments, it is increasingly being recognized that environmental and economic safeguards have to be built into programs for inducing development to extend the long-term sustainability of this development. Indeed, since these programs are man-made, they must contain the possibility of incorporating not only incentives to growth, but also safeguards against the possible adverse effects of growth. They ought to be designed in such a manner as to address all the key dimensions of lasting development. When development is induced and accelerated through planned government interventions, it is essential to ensure that such interventions lead towards durable, self-sustaining, long-term development—not simply cause what we can call brief and ephemeral development spurts.

Agricultural development through purposive programs is probably one of the clearest instances in which sustainable development can and must be achieved through balancing the design of development interventions so that they address, simultaneously, the economic, sociocultural, environmental, and technical dimensions of development. The conference was dedicated to Egypt's agricultural development and I choose examples for my argument from a domain critical for agricultural development—and of high priority in Egypt: development of irrigated agriculture through government programs.

What does it mean to ensure the cultural and social sustainability of the programs for irrigation development? First, I will refer briefly to the experience we have accumulated in the World Bank in designing and financing irrigation projects. The World Bank has a long tradition of supporting irrigation, by financing primarily the physical infrastructure of irrigation systems. However, with physical infrastructure only, that is, without the institutional arrangements to manage such systems, sustainable irrigation development cannot take place. But now it is necessary to define the term *institutional arrangements*. Most often, irrigation programs have dealt with such institutional issues by designing government administrative bureaucracies to manage the systems. Much less attention has been paid to the creation of institutions at the grassroots level. And this oversight—which has been chronic in the design of many irrigation programs—is an oversight of the *components*—the important social and cultural components—necessary for sustainable irrigation development.

Indeed, while top-heavy irrigation bureaucracies have expanded, multiplied, and flourished, the organization of a network of water user associations (irrigator societies) has lagged far behind. Programs have not incorporated in their packages the provisions necessary to create stable, durable, and enduring patterns of social organization at the level of the water users themselves. Thus, these water users—the most important social actors in any agricultural development—have not been organized adequately to carry out their collective tasks—tasks inherent in the use and maintenance of medium- or large-scale irrigation systems. The absence of such organizations has undermined many a program and has led to the early deterioration, and even collapse, of the physical structures of the systems.

Users must be organized collectively to participate in the management and maintenance of the systems, together with the administrative authorities. Absence of water user organizations results not only in the absence of technical maintenance activities, but also in the absence of a "culture of maintenance." The culture of maintenance is the totality of attitudes, stimuli, rules, enforcement mechanisms, rewards, and sanctions that can influence the users' behavior toward protecting the system, making better use of water resources, penalizing abusers, and rejecting destructive practices.

Irrigation programs are not the only development projects to neglect the social structures of sustainability. Similar situations occur in other programs and strategies and any conclusions must be formulated in a broader manner, namely, referring to the importance of organization building as a strategic resource for development. Two concepts—mentioned earlier—have direct relevance: organizational intensity and organizational density. I referred to these concepts in a paper that I wrote a few years ago, "Farmer Organizations and Institution Building for Sustainable Development" (Cernea 1987), and I would like to elaborate them here and suggest them for broader examination.

Organizational Intensity

What I propose first, is that to accumulate the building blocks of social and cultural sustainability, we need to design organization-intensive development strategies, not only technology-intensive strategies. Development aid agencies have long been concerned with technology transfer and have emphasized technology-intensive aid strategies. Many developing countries, inspired by this model, have formulated their own plans as technology-intensive plans. But technology, no matter how advanced, cannot actualize its full potential unless embedded within adequate patterns of social organization—patterns apt to sustain, use, and maintain it. Creating and strengthening such socio-organizational patterns is no less important than the technology itself. Indeed, if organization is seen to be a strategic resource for development, if it enhances human potential, mobilizing people toward collective actions, maximizing synergy, then building up the levels of organization in society is a direct way to increase the effectiveness of development strategies.

The organizational intensity of a development strategy is the degree of presence or absence of provisions for building organizational capacity into that program or strategy. When development alternatives are being contemplated in terms of likely sustainability, it is legitimate to ask, What is the degree of organizational intensity of this, or that, strategy? To what extent does a particular development package include provisions for organizational build-up? Various investments options are always possible and they are likely to have organizational demands of various intensities. These demands need to be met.

There are no rigorous yardsticks for measuring this organization intensity. Often, qualitative analytical judgments must be made to assess the need for organization creation, change, and development. The absence of precise measurements, I admit, is a difficulty. Possibly the science of organizations will eventually develop such measurements. But the fact that we do not yet have quantified means to measure precisely the organizational content and intensity of a strategy does not justify the substantive neglect of the need to organize. The lessons of many evaluation studies underscore the strong association that exists between economic sustainability and organizational/institutional building. We must conclude that it is essential to examine each development strategy in terms of its organizational intensity and organizational content at the time of its formulation. The intensity of organizational content should be regarded as an effective and acceptable indicator among those used to judge ex-ante the suitability and feasibility of a given development strategy.

Increasing the organizational intensity of development strategies usually requires additional investments. It is appropriate to say that not enough is known about how to invest in human organization. It is by far easier to spend vast amounts of money on physical items than to invest effectively and profitably in organization. Yet the returns from higher degrees of adequate organization (as opposed to cumbersome, inadequate organization) are usually very high.

Organizational Density

The other concept, organizational density, is closely related to organizational intensity. Organization-intense development strategies result in "constructed social environments"—to use the concept proposed by Coleman (1990)—that are organizationally "dense." Organizational density reflects the frequency with which types and forms of organization occur in a given social environment, and the multiple belongingness of individuals to organized forms of social action. The organizations themselves may span a wide spectrum, from informal to formal. Sociological studies have demonstrated the high power vested in informal patterns of organization, which sometimes are even more powerful than formal organizations. Variations in the degree of formality or informality, of structure or lack of structure, are not necessarily predictors of the amount of influence which the organization can exercise.

There is significant evidence from the performance of various agricultural production systems that high organizational density can increase returns to producers. Multiple organizational membership of farmers in organizations is a characteristic which varies from one cultural setting to another. For instance, the organizational density of rural societies in Thailand and the Republic of Korea is considerably higher than that in the rural societies in Tanzania, India, or Senegal. Density can be measured through some simple measure like multiple membership or number of organizations. Studies available thus far confirm the correlation and bolster the conviction that organizationally intensive development strategies are superior.

Values and Norms

Another cultural dimension of sustainability in agricultural development—and in development in general—that I want to mention is the role of values and norms and the relationship between organization and norm enforcement. By norms as part of a culture I refer to the entire variety of values and rules, ranging from the informal to the more formal ones, from customary to legal rules, which pertain to economic activities. Human activities are infused with norms, and people at every step either obey or transgress them. To what extent are certain norms more conducive to sustainable development activities? Clearly, those norms that are protective of the environment, that support thriftiness and honesty, defend property, assert the value of mutual support, and so forth—these are necessarily conducive to more effective and sustainable production activities. Increasing the cultural and social sustainability of development programs implies, therefore, the purposive cultivation of attitudes consistent with such norms: in other words, the socialization of the actors, more specifically of the productive agents, towards a normatively regulated, sound economic behavior.

I assume that even those economists who may be little concerned with cultural sustainability would nevertheless recognize the importance of *legal frameworks* for the functioning of free markets. But again, in any society the legal frameworks are a constituent part of what sociologists and anthropologists call culture. Indeed, what else are legal frameworks than cultural norms that are institutionalized by the state? I believe, therefore, that we can agree about that, even with the most hard-nosed economists, and argue jointly that the sociocultural sustainability of induced development is not less important than the economic or environmental sustainability.

References

Cernea, Michael M. "Farmer Organizations and Institution Building for Sustainable Development." World Bank Reprint Series No. 414. Reprinted with permission from *Regional Development Dialogue,* 8:2, Summer 1987, 1–24.

———. "Modernization and Development Potential of Traditional Grass Roots Peasant Organizations." World Bank Reprint Series No. 215. Reprinted with permission from *Directions of Change: Modernization Theory, Research, and Realities.* Edited by Mustafa O. Attir, Burkart Holzner, and Zdenek Suda. Boulder, Colorado: Westview Press, 1981.

Coleman, James. *The Foundations of Social Theory.* Cambridge and London: Belknap Press, 1990.

Daly, Herman E., et al. "Operationalizing 'Sustainable Development' in the World Bank." Working draft. Washington, D.C., World Bank, 1992.

El-Ashry, Mohamed T. "The World Bank, Global Environment, and Sustainable Development." Tokyo, Japan, March 1991.

Tomich, Thomas P. "Sustaining Agricultural Development in Harsh Environments: Insights from Private Land Reclamation in Egypt," *World Development,* 20:2, February 1992, 261–274.

16

Small Farmer Households and Agricultural Sustainability in Egypt

Nicholas S. Hopkins

This paper argues that any concern for sustainability of Egyptian agriculture has to bear in mind the central role of small farmers (Springborg 1990). A socially sustainable agriculture should be one which is built around the dynamism, energy, and ingenuity of the small farmers. The need is to combine a system of production through the small farm household with an increasingly free and open market. At the same time, it is important to be clearminded about the present and probable future role of these small farmers. To stress the role of small farmers is to underline the importance of the family and the household as key organizational units in agriculture and rural life. The household is not only important as a managerial unit, but also as the institution in which the reproduction of the society occurs. The household is the link between the labor process in agriculture on the one hand and family and social continuity on the other. That this has political implications was already sensed by Sayed Marei (1957: 243) nearly forty years ago when he wrote that: "A peasant who owns the land he cultivates (self-supporting) is considered a sound basis for a democratic society."

But if a society of self-sufficient small farmers is suited to democracy, are these farmers also productive and efficient? It can be argued that the present structure of Egyptian agriculture, dominated by small farms, has maintained consistently high yield-levels, despite bureaucratic handicaps. The economist Simon Commander (1987: 227–229), after considering evidence from a survey in 1984, argued that "No consistent trend in terms of productivity was found to exist across farm size class," though larger farms often had higher crop values per unit of land reflecting their stronger position in the market. He infers that if larger farmers have higher yields it is because they have better land, presumably because they have invested more in it, and suggests that small farmers do better with their relatively poorer factor endowments than larger farmers do with theirs (Commander 1987: 176).

Small farmers in Egypt make good use of mechanization, for irrigation, plowing, threshing, transport, and other tasks. They are integrated into complex labor markets. Practically every household either hires in labor, or has members who hire themselves out for wages to farmers, construction contractors, and so forth. The small farmer household is overwhelmingly committed to the market: all farmers sell at least some of their crops, and all householders buy at least some of what they consume. Probably the small farmers are at somewhat of a disadvantage in some of these dealings because each one acts as an independent unit rather than combining with comparable farmers to seek "economies of scale." Thus small farmers are linked vertically to machine owners, labor contractors, and market middlemen, rather than horizontally to each other.

The Egyptian village is not a tabula rasa on to which one can impress whatever one's ideology prefers. In a village are individuals seeking to achieve their goals, incorporated into family structures and organized in households; these households in turn are part of the village structure. The choices that individuals make are structured in ways that make sense to them. There is nothing inevitable about the present structure, but for the moment it guides their choices. Some of this structure, and some of these choices, can best be understood with reference to the national economy of Egypt and the integration of that economy and polity into an international division of labor—the world market. Yet these choices are not made by hypothetical (economics) men and women, responding to economic and financial incentives in disregard of social values and cultural concerns. Rather, they represent the outcome of people operating in an institutional (social and cultural) context.

These farmers are active in seeking their goals—such as raising income to allow for family continuity—while remaining responsive to the state's plans for agriculture through the cooperative and village bank systems. Farm households follow mixed strategies, including not only farming but also local day-labor, labor migration, education, government employment, ownership and rental of machinery, and trade. The role of women differs by region and class, and is changing under diverse outside influences. The household is the basic economic unit, the locus of exchange deriving from the gender-based division of labor; households are then related to one another in communities (reflecting community values and processes) and through the market (buying and selling labor and other commodities). Some patterns of cooperation at the local level, such as the de facto (informal) irrigation water user associations that all farmers adhere to, are also significant.

The Farm Household

I distinguish between the house (*manzil* or *maskan*), the household (*bayt*), and the family (*usra* or *aila*). The household is an economic unit based on people who in various ways live, produce, and consume together. Through its individual members it may have rights to certain assets—land, animals, jobs, income. The household is almost invariably built on relations of family or kinship, but the ideology of kinship should be distinguished from the economic unit. The economic unit contains within it the basic division of labor, based on gender and age. The gender-based division of labor is structurally more significant than the age differences. In fact, the gender-based division of labor operates primarily at the household level, where it is reinforced by conventional understanding of what the roles involve. This means that in most cases husbands and wives do not have to negotiate what their roles are, but can simply adopt the model provided by the culture. This model can be shattered by new patterns, such as those brought home by migrants returning from countries where different but prestigious models prevail (Abaza 1987).

Households in rural Egypt are built predominantly around a nuclear family, parents and children, to whom may be added more children, and perhaps a surviving older-generation parent or other relative. Husbands with two wives can be included here as well. This could be called the nuclear family (*usra*) household. Some households are based on an extended family (*aila*) household, typically when the sons marry and remain within the same household, or when two or more married brothers continue to form a single household. Thus a practical definition of an extended family household is one that contains two or more married couples. One of the understandings of a family, especially in the rural setting, is that it has continuity from one generation to another, and the extended family household is certainly seen as a realization of this model.

Egyptian census figures appear to be based on the notion of the family (*usra*) rather than the household (*bayt*). My impression is that censustakers and others count each married couple as a separate family regardless of the living and working arrangements. One of the consequences of this is that figures we have compiled from our own work of the size of households are larger than the official government figures for the size of family. One can generalize that the typical household size in rural Egypt is around seven persons (Hopkins et al. 1980: 34–37). This has clear implications for the ability of the household to provide its own labor. The average family size is lower: the 1976 census estimated the average size of a rural family in Upper Egypt at 5.0 persons and in Lower Egypt at 5.6 persons. This difference can be taken as illustrative of the size difference between "household" and "family," and also of the contrast between Upper and Lower Egypt.

The Role of the State

Although currently the state in Egypt is seeking to withdraw from some involvement in "civil society," for a generation it was not so silent a partner in agriculture, working through the network of cooperatives and village banks. It provided much of the infrastructure, including irrigation, credit, inputs, and market and storage facilities; guaranteed and enforced the land tenure laws; sponsored the cooperatives; and located the country within the international market situation by seeking market outlets for agricultural products. However, Adams (1986) argues that the state's role has been to manage and control rather than to seek development. In this sense the major push for development has come from the efforts of individual farmers of all sizes. The state's intervention has often been structural and distant, such as through redoing the engineering of irrigation, and not in the details of local social organization.

At present there is a trend to restrict the role of the state, particularly by privatizing the supply of inputs (including credit) and the marketing of crops. How this reduction of the role of the state will work out at local level is not yet known. In mid-1992 a revised land tenure bill was passed in parliament. Although details of its implementation are not yet clear, it seems to signal the state's withdrawal from its guarantee of tenure rights in rented land. This clearly has the intention of fostering land consolidation, thus eliminating many small farmers, or at least reducing the size of their holdings (Springborg 1991).

Land Tenure

The general picture is one of individual land tenure; transfers of title take place through inheritance and the market. The operative farm unit is the holding, defined for the last generation as: land owned plus land legally "rented in" minus land legally "rented out." Most often, the part that was "rented in" occupied between one-third and one-half of the holding. After the 1961 land reform, this holding was recorded at the cooperative and was the basis for dealings between the cooperative and the landholder. There is also substantial unofficial rental, at a much higher price, reflecting market conditions. This is not recorded officially.

Most holdings are small; the national average is around 2.5 feddans. There are also medium and large holdings, although the very largest—with thousands of feddans—were expropriated in various ways in the 1952 and 1961 land reforms. After 1969, the maximum holding in principle was 50 feddans, but a few landholders farmed several hundred feddans. There are many "landless" in the rural areas. Though some work as laborers, not all are involved in the agrarian structure: many—civil servants, merchants, commuting factory workers—have other occupations.

Differented access to land has long been a major feature of the political economy of rural Egypt. I detect a slight trend towards a new concentration of landownership, especially in the new lands, but it is less significant than the concentration of ownership of agricultural machinery, or the ability of the larger farmers to take advantage of new market opportunities. Any trend towards concentration of landownership is still inhibited by the agrarian reform rules that limited the size of holdings. For the time being, the predominant pattern in Egyptian agriculture is one of many small farmers. The relatively few larger ones are often better able to use their land intensively because of their greater access to capital and markets.

Technology and Capital

Egyptian agriculture is fairly heavily mechanized. Many tasks are done with tractors, but others, such as planting, weeding, and harvesting, are still largely carried out by hand. We are not looking at mechanized crops, or even farms, but at mechanized *tasks*. Tractors are used to plow the land, haul wagons, run threshing machines, and some other tasks. Pumps are used to lift water for irrigation.

There is little difference in the use of mechanization between large and small farmers; the difference is in the ownership of the machinery. In a 1982 sample (Hopkins et al. 1982), 90 percent told us they used tractors for plowing, 85 percent for threshing, and 15 percent for transport from field to village; 62 percent used irrigation pumps. But hardly any reported using machines for harvesting, planting, or weeding.

Few farmers own their own machines, especially tractors; instead they rely on the rental market. Tractors available for rental typically—in 90 percent of cases—belong to large farmers. The cooperative provides very few tractors—one or two per village; sometimes none at all. The market for rental machines in each village tends to be highly concentrated: the proportion of farmers who reported using the main renter—that is, the biggest—in each of ten villages averaged 37.3 percent, with a range of 13 percent to 53 percent. People generally say they use machinery "to save time and effort"; those few (18 percent) who preferred animal and human power argued that it resulted in higher quality production. We found a positive correlation between the use of machinery and hired labor, suggesting that those who hire machinery also hire labor. However, hired labor is very common everywhere. Somewhat paradoxically, when people talk about saving time and effort, they partly mean the time and effort spent supervising labor, which is greater under nonmechanized conditions.

No account of Egyptian agriculture is complete without reference to the role of animals. Almost all farm households try to keep a milk animal and perhaps draft animals (Hopkins et al. 1980; Zimmermann 1982). Caring for the animals in the stalls (including milk and the preparation of dairy products) is usually a woman's chore. Women also have sole responsibility for fowl and rabbits in the house. Chickens raised for meat or eggs on an industrial scale are usually tended by men (Saunders and Mehenna 1986). Men raise berseem, which is often cut and taken to the animals in their stall; men also take charge of buying and selling animals.

Labor Process

The labor process is fragmented. The analytical starting point is the commodity-producing household operating in the context of the village or other rural community and in conjunction with the state. The head of a rural household is primarily a manager, linked to other households through exchanges and hiring of labor, rental of machinery and land, and other relationships. Mechanization has contributed to the creation of a labor market by breaking up the household labor process, by fragmenting it so that households must have recourse to hired labor. Each task in the sequence of growing a crop requires a different combination of machinery, money, labor, water, etc., and farmers typically grow four or five different crops.

In our 1982 sample (Hopkins et al. 1982), farming was listed as main occupation by 81 percent of the group; 10.5 percent followed nonagricultural occupations; and the remainder were widows, pensioners, etc. (See also Radwan and Lee 1986; Commander 1987.) There is almost never an exact correspondence between household composition and the labor requirements of the farm unit. Nowadays, one of the most common features of agriculture, in Europe and North America as well as Egypt, is the part-time farmer. Some household heads work outside agriculture; in other cases, one man works on the farm while others work elsewhere, and income is to some degree shared. In the villages near Bilbeis, for instance, many men work in the Tenth of Ramadan, about 40 km away (Hopkins and Hamdy 1990).

The decentralization of the hiring of labor by the household minimizes the need for hierarchical control of the labor process. The small size and shifting composition of work groups favor the ability of the household head to make micro adjustments and decisions. One of the tasks of the household head is to manage the labor input in agriculture: either that labor is from the household and subject to discipline in terms of family values and norms, or it is hired labor, subject to community values and norms. For the most part the number hired and managed by each unit is small: control of labor is thus intensive. Only the household head follows the crop through the year (crop cycle), though others do much of the physical work. With this division of labor comes a concentration of knowledge and hence something of a deskilling of farming.

Labor migration has always been an important household strategy in rural Egypt (Hopkins et al. 1982; Adams 1991; Brink 1991; Nada 1991).

The new element is that people are migrating to outside Egypt, which makes them subject to radically different conditions and values (often conservative in social terms). The larger amount of money people can thus earn means that this is not just a survival strategy but a chance to make a quantum leap upwards in material terms. The return of such migrants also provides another symbol of the involvement of Egyptian villages in the world economy and culture. There is no question that the money brought back and sent back by migrants has contributed to an improvement of living standards in the countryside, most visibly through an upgrading of housing, but also through the financing of nonfarm entrepreneurial activities.

Women work at home, care for animals, organize consumption. In the Delta they also trade at local markets and work in the fields—either family fields or, occasionally, for wages (Zimmermann 1982; Saunders and Mehenna 1986; Abaza 1987; Toth 1991). In Upper Egypt, since women do not work outside the house, and so not in agriculture, it places more strain on the men to accomplish all the tasks and generate the streams of income necessary (Hopkins 1991a). Where women work for wages, as in the Delta, they are typically paid about 60 percent of the male wage, and are restricted to certain jobs. Thus it is not clear that women and men are interchangeable. Some employers (near Bilbeis) report finding that women are preferable because they are more "obedient." Working on one's own farm is more common, and from one village in Minufiyya it was reported that a woman's farming skill is taken into account when deciding on a marriage (Glavanis and Glavanis 1983).

The Labor Market

In the labor market, there is limited interchangeability between men and women, adults and young, because they are usually assigned to different tasks. The big farmers set the wage scale: among the techniques used to lower the general wage level is the ploy of hiring work gangs from neighboring villages where there may be more of a labor surplus. There is a complex wage structure: different tasks are paid in different ways. The control of labor is diffused, and the decentralized control of labor is reinforced by the fact that the worker must constantly be rehired. Often the farmer hiring labor does so through a contractor. This simplifies his task and the task of labor discipline devolves onto the labor contractor. Even those who are de facto permanent workers say (as do their employers) that they are not permanent.

Going to Market

Since the household is the basic production unit, it also takes charge of the marketing. Some crops are sold to the government, such as cotton, lentils, and soybeans. Others are sold to private merchants, such as some cereals, fruits, and vegetables. The cereal crops and some other secondary products (straw for fodder, stalks for fuel) are sold separately. Fruits and vegetables are usually sold to agents of the big dealers in the Cairo and Alexandria markets, though some go to dealers from Asyut and other towns, or are sold directly to consumers. In many cases the merchant or agent is the farmer's fellow villager. Farmers attempt to deal with and control this situation, but its complexity is often beyond them. Individual small farmers do not pull much weight in this market situation: they are "price takers."

In general, the farmer deals with a different set of merchants and a different procedure for each crop. He is usually paid for the crop sometime after delivery, usually after the agent or merchant has resold the crop. This adds to the complexity. Merchants and their agents will sometimes make an advance to the farmer, financing the crop. This amounts to dealers maintaining control over farmers through indebtedness. Farmers who can remain free of this system and sell their crop at harvest to the highest bidder generally do better. Farmers are likely to retain part of their cereal crops for their own use. In fact, a majority of farmers in our study said they did not sell any of their wheat, berseem, or corn. This probably reflects their success at growing only the amount that they expect to consume. In other words, this situation does not indicate a "subsistence" orientation, but rather its opposite. They will grow a food crop for home use only if that is cheaper than buying it.

Village Politics

The village in Egypt is still a basic administrative structure, with institutions and processes. Villages are becoming more urbanized, but many also preserve vigorous local institutions of self-government that act as a screen between the designs of the state and the life of their residents. This self-government is based on the local power of the village elite. The village is also the locus where the relationships among households are worked out. Here is where the landless and the landholders relate to one another, where the large farmers set the pattern which the small only aspire to, and so on. Thus it is at the level of the village that first of all one can discern the class situations, the social contexts in which class roles are learned and acted out. The classes themselves are an outcome of the mode of production. From a national perspective, elements are absent from the village, such as the bourgeoisie, the truly large and capitalist farmers who reside in the cities, and the elite of the state bureaucracy. Thus the class structure in the village cannot be analyzed in isolation from that of the nation.

Changing Relationships

At present in the villages, patron-client relations, linking the rich and prestigious to the poor and anonymous, are more evident than class relations (Hopkins 1987: 157–177; Hopkins 1991a). Something of the sort was apparently emerging at the time of Gamal Abdel Nasser, but it seems that it went into remission in the early 1970s. The relationship between households is often unequal. These differences are both real (material) and symbolic. Among the symbolic differences are various signs of deference (gifts of cigarettes, evening gatherings, and directionality of hospitality). But the social organization of agriculture, insofar as the hiring of labor and machines takes place between households, reinforces unequal relations between them. In sum, the patron-client relations mark the inter-household relations.

In the community are set the values by which economic activity is regulated. There is a sense of fairness, of right and wrong. Negotiations over power (wages) are not only carried out face to face, but through community processes, such as the intermittent debate on what is fair, what is going on elsewhere in Egypt, what is necessary, and so on. The same applies to questions such as women's proper role. The settlement of disputes is also a community process; so is popular religion. The spread of education represents in part the ability of communities to organize themselves to seek new resources: frequently new schools are built with contributions from villagers, especially the rich. This is perhaps an extension of the continuing pattern of support for mosques and other religious foundations.

Another level of integration that is important for the sustainability of Egyptian agriculture, and the small farmer within it, is the region. Villages, unequal in the opportunities that they offer, often specialize. For instance, in Tukh Markaz in Qalyubiyya governorate, the village of Ammar specializes in apricots; the village of Deir in strawberries; and the villages around Kafr Mansour in citrus and citrus marketing. Namul features eggplants; Imyay the manufacture of crates for transporting fruits and vegetables; and another cluster of villages manufactures charcoal for the Cairo market. Other villages in the region grow berseem for sale to villages whose land is devoted to one of the other crops. Similar patterns could doubtless be found in many other areas (Hopkins 1987: 30–33). These factors should be in mind when thinking about the market relationships of the small farmers.

The Future of Rural Life

One has to visualize the Egyptian small farmer not as a peasant but a petty capitalist who is active simultaneously in the labor market, the machine rental market, and of course in the purchase of inputs and sale of outputs. The difference between the large and small farmer is thus one of degree:

they do not represent, for instance, different "modes of production." One might wonder whether the basic smallholder pattern that has dominated in Egypt since the agrarian reforms of the 1952–1961 period will endure much longer. What will be the shape of Egyptian agriculture, the country-side, or the farm household, in the twenty-first century? As members of the farm household get more and more involved in education, off-farm employment, labor migration, and the like, the match between the house-hold and the labor requirements of the farm will be less evident. There are also pressures to eliminate fully the role of the state in marketing and input-provision, which would undercut the cooperative's function for the last generation. And there is pressure to extend the market in land.

Given the general structure of Egyptian society, there is no reason to fear that the household and the family will soon disappear, despite these pressures. But these institutions could certainly change their functions. The last forty years have shown the value of the small farmer household—the household of petty commodity producers—in providing a thread of continuity in rural Egypt. Agrarian reform—that directed Egyptian agri-culture away from a structure based on large estates—proved to be successful in creating the structures which now seem likely to change under the pressure of "structural adjustment" and the market principle.

Should the Egyptian policy continue to support the small farmer, and the corresponding family household pattern of social organization? Will the pattern continue, whether supported or not? If the goal is a sustainable agriculture—that is, growth with balance, without using up tomorrow's resources today—then what role does the family household have to play? Can we assume that the family-based farm is more conducive to a sustain-able agriculture than is the large-scale, fully capitalist one? Before such choices or decisions can be made, adequate social and cultural under-standing of the current functions and structure of the household is needed. A failure to understand rural social organization can lead to policies that destroy the fabric of rural life—as has happened in Iran (Hooglund 1982), Iraq (Batatu 1978), Algeria (Marouf 1981; Chaulet 1987), and other countries. And the destruction of rural life can lead to accelerated migra-tion of rural people to urban areas, and a rootless urban class that can provide the social basis for a new regime. Not only agricultural sustainabil-ity but political stability may be at stake!

References

Abaza, Mona. "The Changing Image of Women in Rural Egypt," in *Cairo Papers in Social Science.* 1987, 10(3).
Adams, Richard H. *Development and Social Change in Rural Egypt.* Syracuse: Syracuse University Press, 1986.

————. *The Effects of International Remittances on Poverty, Inequality, and Development in Rural Egypt.* Washington, D.C.: International Food Policy Research Institute, 1991.

Batatu, Hanna. *The Old Social Classes and the Revolutionary Movements of Iraq.* Princeton: Princeton University Press, 1978.

Brink, Judy H. "The Effect of Emigration of Husbands on the Status of Their Wives: An Egyptian Case," *IJMES*, 1991, 23(2): 201–211.

Chaulet, Claudine. *La Terre, Les Freres, Et L'Argent.* Algiers: Office des Publications Universitaires, 1987.

Commander, Simon. *The State and Agricultural Development in Egypt Since 1975.* London: Ithaca Press, 1987.

Glavanis, Kathy R.G., and Pandeli Glavanis. "The Sociology of Agrarian Relations in the Middle East: The Persistence of Household Production," *Current Sociology*, 1983, 31(2): 1–109.

Hooglund, Eric. *Land and Revolution in Iran: 1960–1980.* Austin: University of Texas Press, 1982.

Hopkins, Nicholas S., et al. "Animal Husbandry and the Household Economy in Two Egyptian Villages," Report to Catholic Relief Services, Cairo, 1980.

Hopkins, Nicholas S. *Agrarian Transformation in Egypt.* Boulder: Westview, 1987.

————. "Women, Work, and Wages in Two Arab Villages," *Eastern Anthropologist*, 1991, 44(2): 103–123.

————. "Clan and Class in Two Arab Villages," in *Peasants and Politics in the Modern Middle East.* Edited by Farhad Kazemi and John Waterbury. Miami: Florida International University Press, 1991a, 252–276.

Hopkins, Nicholas S., Sohair Mehenna, and Bahgat Abdelmaksoud. "The State of Agricultural Mechanization in Egypt. Results of a Survey: 1982." Report to the Ministry of Agriculture, Cairo, 1982.

Hopkins, Nicholas S., and Iman Hamdy. "Social Issues in Agricultural Mechanization." Report to the GTZ, Cairo, 1990.

Marei, Sayed. *Agrarian Reform in Egypt.* Cairo: Imprimerie del'Institut Français d'Archeologie Orientale, 1957.

Marouf, Nadir. *Terroirs Et Villages Algeriens.* Algiers: Office des Publications Universitaires, 1981.

Nada, Atef Hanna. "Impact of Temporary International Migration on Rural Egypt," *Cairo Papers in Social Science*, 1991, 14(3).

Radwan, Samir, and Eddy Lee. *Agrarian Change in Egypt: An Anatomy of Rural Poverty.* London: Croom Helm, 1986.

Saunders, Lucie W., and Sohair Mehenna. "Village Entrepreneurs: An Egyptian Case," *Ethnology*, 1986, 25(1): 75–88.

Springborg, Robert. "Rolling Back Egypt's Agrarian Reform," *Middle East Report No. 166*, 1990, 38: 28–30.

————. "State-Society Relations in Egypt: The Debate Over Owner-Tenant Relations," *Middle East Journal*, 1991, 45(2): 232–249.

Toth, James. "Pride, Purdah, or Paychecks: What Maintains the Gender Division of Labor in Rural Egypt?" *IJMES*, 1991, 23(2): 213–236.

Zimmermann, Sonja D. *The Woman of Kafr el Bahr.* Leiden: Research Center for Women and Development, 1982.

17

Women's Rights as a Condition for Sustainability of Agriculture

Hoda Badran

The concept of sustainable development has become part of the vocabulary of development since the publication of *Our Common Future*, the 1987 report of the UN World Commission on Environment and Development under the chairmanship of Gro Harlem Bruntland (World Commission 1987). In this report, sustainable development is defined as "development which meets the needs of the present without compromising the ability of future generations to meet their own needs." Other definitions of the concept have been proposed, but there is a general agreement that sustainable development essentially implies (1) a balance between meeting the needs of the present and future generations; (2) prudent management of available resources and environmental capacities and the rehabilitation of the environment previously subjected to degradation and misuse; (3) implementing preventative policies to promote environmentally sound development; and (4) adopting objectives that would include: achieving growth with quality; addressing the problem of poverty and satisfaction of human needs; handling the problem of population; adapting science and technology; and ensuring participation in human interaction with the environment and the economy (FAO 1991).

The concept of sustainable development has at its core the issue of justice and fairness, particularly towards future generations. There is a time dimension here. Development is set on a continuum; and today's actions are seen as an extension of yesterday's choices, resulting in the type of world to be inherited, tomorrow, by our children. There is a time dimension also in terms of urgency—of making difficult but necessary choices to effect change immediately to slow down, or prevent, environmental degradation. This time dimension links the present with the past and future; gives a history to the situation; and highlights how economic policies and development strategies undertaken in the past have affected the outcomes today. It is unfortunate that deterioration of the human environment due to scientific and technical developments has been generally regarded as an inevitable by-product. It was not at all seen as an interference with men's or women's rights to a standard of living adequate for full family health. Nobody is blameless for the present terrible situation. Each of us should be not only aware of the problem but share the responsibility in bringing about the necessary changes.

Women in the Development Process

Notwithstanding the accelerated rate of urbanization that Egypt has witnessed since the 1960s, the agricultural sector still absorbs 38 percent of the labor force, and its output contributes 21 percent of the GDP. Agriculture plays a key role in the Egyptian economy in many ways: it is a major source of income and a way of life for a sizable part of the population. Sustained economic growth in Egypt is closely linked to the prospects of growth in the agricultural sector on a sustainable basis. However, this is constrained by several factors, some related to the size and quality of arable land, others to water resources, technology, and the macroeconomic environment.

To alleviate these constraints, the agricultural strategy is part of a larger strategy to shift Egypt's economy from unsustainable growth, characterized by dependence on services and nontradables, to sustainable growth, led by the traded goods with an expanded role for the private sector (Richards 1991). This implies a long-term perspective that creates a balance between the needs of existing and future generations; between the needs of the different socioeconomic classes; and between the needs of men and women. Such balance cannot easily be achieved, but it is essential if the new Egyptian strategy is to succeed. In fact the national strategy has to take into account three distinct but interacting dimensions of agricultural development: (1) technical/physical; (2) financial/economic; and (3) human/institutional. This reflects the fact that agricultural sustainability is not the business of one discipline. It is a multiple responsibility with each discipline having its range of activities that follows from the tools and language it uses.

The environmentalists, for example, when talking about the physical dimension, mention land-use, dissemination of technology, and so forth. Agronomists talk about credit markets, agricultural technology, and the like. Economists discuss costs, prices, subsidies, and maximizing returns. Those in the field of humanities use yet other words to discuss human and institutional issues such as knowledge, skills, leadership training, health, communication, management, and—most important—gender equity, or the role of women (Woods 1987).

Fortunately, there is an increasing recognition of the human dimension—an understanding that the engine of agricultural development and sustainability is people, and not merely the natural and physical resources, as was believed by some. Fortunately, the world has discovered that rural women have a central role in achieving and maintaining agricultural sustainability. There has been no recognition, though, that they were the first to comprehend the core issue in the concept of sustainability, i.e., justice and concern for future generations. Whether by instinct or because of intelligence, they were always, before the concept even came into the open, able to compromise their own needs for the future benefit of their children . Studies have documented that women, more than men, give priority to the satisfaction of their children's needs out of their own income.

Environmental Managers

The literature on agricultural sustainability took note of women only when it was recognized that they were to a great extent the managers of environmental resources. They are the ones who fetch water and use it in preparation of food or in washing. They are the ones to collect and prepare fuel, and use it in cooking and baking bread. They raise the cattle and look after the animals. They also play an important role, at least seasonally, in the crop production process itself. They are occasionally recognized as producers absent from the labor statistics, but they are perceived essentially as reproducers. Sometimes they are accused of polluting the environment because they use the water resources for bathing and washing, and they fill the air with smoke because they cook with biomass fuel. They are blamed as if they have alternatives to choose from, and as if they are not the first to suffer from the pollution. Literature has until recently discussed rural women as a subsidiary issue and as a means to the achievement of other goals. They should, for example, bear fewer children and hence limit the size of the population, if this is a government policy. They should change food habits, if a new crop is to be introduced; and they have to breastfeed their children, regardless of their nutritional status. In other words, and using research terminology, they are treated as independent and not dependent variables. They are often treated sectorally and not wholly.

I am thankful for the United Nations Organization's raising of the principle of advancement of women, first as part of its charter and of several conventions, then by the Declaration on the Elimination of Discrimination Against Women, proclaimed in 1976. The Preamble of the Declaration of 1976 states:

> Despite the Charter of the United Nations, the Universal Declaration of Human Rights, the International Convention on Human Rights and other instruments of the United Nations and of the specialized agencies, and despite the progress made in the matter of equality of rights, there

continues to exist considerable discrimination against women.

The preamble points out that "discrimination against women is incompatible with human dignity and with the welfare of the family and of society, prevents their participation on equal terms with men in the political, social, economic, and cultural life of their countries, and is an obstacle to the full development of the potentialities of women in the service of their countries and of humanity" (United Nations 1984). The Convention on the Elimination of All Forms of Discrimination Against Women, which was adopted on December 18, 1979, is a major step forward in recognizing women in their own right. It includes a special part on rural women (Article 14) as a recognition of the particular problems faced by them and the significant role they play in the economic survival of their families, including their work in the nonmonetized sectors of the economy.

Without going into the details of Article 14 of the UN Convention (1984), and taking into consideration other articles, the rural woman can achieve her basic right for a healthy status, and hence be part of a sustainable agricultural sector, if she can attain access to:

- Adequate health facilities and living conditions
- Economic opportunities, including credit, technology, etc.
- Education, information, and training
- Freedom to organize, to make choices, and participate in community decisions

It is, therefore, appropriate to address the status of the Egyptian women in the agricultural sector as a human rights issue. This approach is most relevant to the concept of sustainability because it puts the well-being of all individuals in focus, which is actually the ultimate goal of development.

The obtaining and development of a healthy life—survival, growth, development, and enjoyment—is a fundamental human right. This status of well-being, which is an output, has to be distinguished from the mere absence of diseases, and from health services, which are only the means (inputs). The two concepts are of course interrelated, but the inputs for a healthy status are several and far beyond the health services. In other words, these services are necessary but not sufficient to produce such status. A healthy status of the individual is actually the outcome of the performance of all of the community delivery systems. Their efficiency and effectiveness in meeting human needs determine the level of well-being of the population.

Women's Health

Women in general, and rural women in particular, are not often healthy in Egypt. They have few choices about their lifestyle and still fewer opportunities to improve their own and their family's status. This is due to the multiple roles they play and the burdens they have to carry. Social, economic, and political factors shape their roles and status and affect their health and well-being, but they have little control over them. How far does the Egyptian rural woman exercise her fundamental right to enjoy a healthy status? I will take some significant indicators to answer the question.

Maternal mortality is an important indicator: it reflects the degree of risk inherent in the woman's reproductive role as well as her health status from early childhood. This risk has been almost eliminated in the industrialized countries. In Egypt, the average maternal mortality is 320 per 100,000 and undoubtedly the risk is much higher in rural areas. In comparison, the average ratio is 27 in Bahrain and 6 in Kuwait (United Nations 1991). How can one rationalize such a high risk of dying to which rural women in Egypt are subjected?

Malnutrition is another indicator, reflecting poverty and unhealthy status. Studies on women seeking family planning devices indicate that in rural Egypt a high percentage of women suffer from anemia—reaching as high as 65 percent in some villages, compared with about 46 percent in urban areas. It is no wonder that, with high incidence of malnourishment among a majority of women, 67 percent of babies have low birthweight (compared to 8 percent for Turkey and 9 for Iraq). The figure for Egypt's rural areas is higher than the average for the rest of the country (Hussain 1988).

Illiteracy is a third indicator. The illiteracy rate for rural women in Egypt is 76 percent, compared with 44 percent for urban women (CAPMAS 1986).

When these three indicators are taken together, the answer to the question is that the rural woman in Egypt has not attained her basic right to a healthy status and cannot be an active partner in a sustainable agricultural sector. A survey of community services available to her should provide an explanation of the situation.

The Rural Health System

The rural health system has increased the number of its health units: there is now one unit per 1.5 villages. The total number of rural health units in 1986 was 2,731 (CAPMAS 1986). The quality of services, however, has been criticized heavily for lack of equipment, insufficiency of drugs, and the attitudes and performance of the health personnel. In a recent study (Hussain 1988), rural housewives stated that they preferred to go to private doctors rather than use the health units. They go to the health unit only for children's immunization: that is the only service they can really get and are satisfied with. Units do not have laboratories and cannot give them the prescribed drugs. Furthermore, the majority of mothers use birth attendants (*davas*), even if they are not licensed. This applies equally to literate and illiterate women.

Although health education is an important function of the health unit, a large number of rural women appeared to be ignorant about types of communicable diseases and the ways in which they may be transmitted. The study was undertaken on a sample of two hundred families only, which makes it difficult to generalize, but it gives an indication of the situation (Third World Forum 1991).

The quality of living conditions is determined to a large extent by the environment—the shelter—in which one lives, and by behavior and life-style. The immediate environment, if not adequate, entails such risks as infectious diseases—malaria, cholera, and tuberculosis. This inadequacy relates to water supply, sanitation, and indoor air pollution. Lack of safe water and sanitation in rural areas affects women disproportionately as they are the principal procurers and users of water. They are often responsible for waste disposal and are at a higher risk of exposure to water- and sanitation-related diseases than are men. In Egypt about 45 percent of rural households are still without access to piped water (United Nations 1991). Rural women have to go to the river or to a public pump to get water to meet their family needs. The water resource may not be near and the water they get is often polluted. Girls go with their mothers and elder sisters to fetch water as soon as they start to walk. The utensils they carry get bigger as they grow older. Fetching water is a part of their life, with no complaint or grumbling, although they may have to undertake this task several times a day. Biomass fuel (obtained at no financial cost) leads to serious respiratory diseases because women are exposed to the smoke that comes from it. Exposure to biomass fuel emissions is probably one of the most significant occupational hazards for rural women.

Education

The data reflecting rural women's exercise of their basic rights to educa-tion and information are disappointing. Although Egypt's constitution gives equal right to education for males and females, and basic education is free and compulsory by law, enrollment figures reflect a serious gap between the sexes, particularly in the rural areas. CAPMAS and UNICEF publications put out in 1989 show that the enrollment ratio for rural girls is 71 percent, compared with 93 percent for urban girls; 95 for rural boys; 96 for urban boys. The gap between boys and girls gets wider in the higher stages of education. School dropout rates for girls are no less discouraging. Quality of education affects both sexes, but stereotyping in textbooks, enforcing prejudice, is particularly serious with regard to females. Literacy classes are rarely attended by rural women. They do not go to these classes even if they are convinced of the usefulness of the learning process, and even if there is no pressure on them not to go from the family.

The direct correlation between literacy, health, economic and political power, and the exercise of informed choice, especially for women, cannot be overemphasized. Education determines the rural woman's access to paid employment, her earning capacity, her overall health, control over fertility, and the education and health of her family. Education also helps women to handle social prejudice and hence be able to participate fully in the community.

In the study mentioned earlier, a high percentage of the housewives, though illiterate, expressed a desire that their daughters go to university if at all possible. This indicates that the obstacle to female education is not a matter of lack of understanding on the part of mothers so much as economic and social factors (Office of the Middle East 1991). As for information and facilities on family planning, Egypt has been using the mass media, particularly TV, to communicate messages and information. There are questions as to whether these messages have been effective or relevant to rural women. There are reasons to believe that many rural women bear more children than they wish to have or know how to prevent. Frequent pregnancies result in childbirth complications, anemia, and diminished working capacity.

All Work and No Play

As to the women's right to economic opportunities, the United Nations Women's Decade has exploded the myth that women in the Third World are peripherally engaged in agricultural labor. For them it is all work and no play. Much of women's agricultural labor—as with domestic work—is overlooked because it is unpaid. Often it does not appear in official statistics. In Egypt, any attempt to estimate the agricultural labor force is complicated by several factors, among which is the treatment of women's labor.

Crop harvests attract large numbers of women and children. There is reason to suspect that the seasonal activities of women in crop production are underestimated. The labor participation rate for women in the agricultural sector in general in estimated to be 16.4 percent, much below the rate in reality. Two surveys conducted in 1978 and 1984 in the Nile Delta indicated that about one-third of all crop labor was carried out by women (Richards 1991). In 1983, for the first time, the labor force sample survey attempted to include the amount of labor done by the family in caring for livestock. The data indicated that about 40 percent of the livestock work was done by women. Since total agricultural labor is divided at a ratio of about 40:60 between crop and livestock labor, the women's livestock labor alone would be about 16 percent of all agricultural labor. Labor statistics in Egypt are not up-to-date. The latest (1983) indicate that about 40 percent of those women who are economically active in the agricultural sector are paid laborers. Women's wages are often lower than those for men and are very close to children's wages, which encourages substitution of female labor for male labor in crop production.

In spite of the contribution of rural women to the agricultural production process, their access and control over productive resources is highly constrained by legal and customary factors. Extension training often excludes women, as if they had no role in production. The new technologies are usually geared towards men, as are credit facilities. A number of international organizations in Egypt are participating in income-generating projects for rural women, trying to improve their access to economic opportunities, but these do not reflect a concrete policy or conviction. I suspect strongly that changes resulting from the structural adjustment policy will make the situation of rural women even worse. Accelerated mechanization may reduce their access to job opportunities. The shift to cash crops may affect the availability of food for the family.

The Right to Choose

Finally, what can be said about women's right to make choices? Life for the rural woman is prescribed from childhood to old age. She does not have the time nor the skills to organize and she is not aware of her right to participate in community decisions. I have tried to touch on several rights which rural women are entitled to. We should confess that much needs to be done to make these rights a reality, and make women part of the strategy to make the agricultural sector sustainable. What needs to be done is not more of the same of what has been going on. Rural women themselves should be allowed to sharpen their awareness of their rights. They should then learn how to increasingly attain them and exercise them. They may be assisted in doing so, but rights are never handed out.

References

CAPMAS (Central Agency for Mobilization and Statistics). *The Census of 1986.* Cairo: 1986.

———. *The Statistical Yearbook, 1988.* Cairo: CAPMAS, 1988.

FAO. "Sustainable Agriculture and Rural Development in the Near East." Netherlands Conference on Agriculture and the Environment. Regional Document No. 4. Rome: 1991.

Hussain, Mohamed Amro. "Proceedings of the National Conference on Social Welfare." (Arabic.) Ministry of Health. Cairo: November 1988.

Office of the Middle East. "Improvement of the Socio-Economic Conditions of the Rural Areas in the Arab Region." Third World Forum. Cairo: 1991.

Richards, Alan. "Agricultural Employment, Wages and Government Policy in Egypt During and After the Oil Boom," in *Employment and Structural Adjustment in Egypt in the 1900s.* Edited by Heba Handoussa and Gillian

Potter. Cairo: American University in Cairo Press, 1991, 25–27.

United Nations. *The United Nations and Human Rights.* New York: 1984, 149–150.

United Nations Development Programme. *Human Development Report 1991.* New York: Oxford University Press, 1991.

Woods, Bernard M. "Human Development and Sustainability in Sustainability Issues in Agricultural Development," in Proceedings of the Seventh Agricultural Sector Symposium. Washington, D.C.: World Bank, May 1987.

World Commission on Environment and Development. *Our Common Future.* New York: Oxford University Press, 1987.

18

Sustainable Development Necessitates a Social Revolution

Greg Spendjian

The global problem about environment and development arises from the confluence and interaction of many problems. To list some of them: resource depletion; environmental degradation; persistent economic problems such as inflation and unemployment; growing gaps between rich and poor, both within countries and between countries; individual alienation and social disintegration; arms buildups; the population explosion; and the escalating use of violence as a conflict resolution response. What we have is a metaproblem—a problem of problems, a global problematique, or as some have called it, "a mess."

In such situations, when you try to solve one set of problems it can frequently lead to a worsening of the situation elsewhere. What is becoming clearer to many, though unfortunately not to those who have the economic and political power, is that the quality of the situation is such that these problems are not amenable to solution individually. This concern for the "environment and development" issue was given an enormous boost, of course, with the convening of the UN World Commission on Environment and Development—the Brundtland Commission—and the publication of their findings (World Commission 1987).

Notwithstanding the continued fuzziness about the notion of sustainable development, there is general agreement that this is something desirable. Can you imagine somebody standing up at a conference and saying, "I do not support the idea of sustainable development." And of course it is clear that decisions continue to be made—at the individual, community, institutional, national, and international levels—which go completely contrary to the notion of sustainability. The reasons for this are very varied, ranging from desperate people attempting simply to secure their survival (be it at the personal, corporate, or national levels) to greed, selfishness, and shortsightedness.

My title is purposefully provocative. After all, Egypt already had its revolution in 1952, but what I am talking about goes well beyond Egypt and the kinds of changes most social revolutions have brought about to date. All the talk about sustainability and sustainable development, the conferences held, books, articles, and reports written, will remain empty and abstract words, nothing but rhetoric, unless people come to accept that the solutions to the problems we face will not be found within the existing political, social, economic, or cultural contexts. There must be a fundamental revolution in the very underpinnings of our societies for us to move in a positive direction.

Sustainable Agriculture Calls for a Sustainable Society

I would like to suggest that sustainable agriculture is unlikely to happen in a society which is not committed to the goals of sustainability *overall,* meaning in all sectors: goals which arise from optimizing at the same time the imperatives of economics, ecology, and ethics or social justice. We know more or less what the immediate requirements for sustainable agriculture are: we must conserve soil fertility; avoid salinization and alkalinization caused by improper irrigation methods; maintain biodiversity; and decrease pesticide use or abuse. Add to these objectives others such as maintaining adequate food security and conserving wildlife. You can then ask, "But how can we do this?" Well, some technological answers are already available: promote use of regenerative production systems which integrate livestock, perennial, and annual crops; promote approaches which use to a much greater extent mulching, crop rotations, intercropping, agroforestry, conservation tillage systems, green manuring; develop and use appropriate integrated pest management regimes; increase irrigation efficiencies and improve drainage systems; diversify genetically away from monocropping; perhaps even reduce the extent of conversion of humanly edible grains to meats (Conway and Barbier 1990). The question then is: How can we bring these and other changes about? And why, if they are so obvious, do they not happen on their own?

Technology Is Only Part of the Answer

We have begun to realize that technological recommendations are not by themselves the only answer. We must also be aware of the social dynamics and the reasons for decision-making by farm families. We also have become more aware that to be effective, or even to be adopted, technologies must fit into supportive policy environments. Fortunately more and more is being written about policy changes which may encourage sustainable behavior: putting in place laws where the polluter pays; internalizing wherever possible the costs of "externalities"; taking into account resource degradation and depletion in the charges made for resource use;

reducing price supports for pesticides and chemical fertilizers; providing incentives for sustainable agricultural production practices; and reviewing land tenure systems, instituting ones which encourage the use of regenerative approaches.

But how can the political will be generated to put such policy changes in place? There are even some answers to that question. Modification of national accounting systems, for example, truly to reflect resource depletion or degradation, in order to build a convincing case for conservation measures. There is also emerging some belated debate on the importance of ecological carrying capacity and the question of "scale." The question of carrying capacity or scale puts us face to face with the total impact of our production and consumption activities on the ecosystems. This puts front and center the extremely difficult and emotive subject of population growth. Anybody who is concerned about the subject of sustainability cannot fail, after a while, to be confronted with the link between carrying capacity and population. However one turns this factor in the sustainability equation, it is impossible to avoid for long the conclusion that at some point populations will have to stabilize if not decrease if we are not to continue the trend toward irreversible environmental damage.

One of the problems with many of the strategies put forward is that they do not proceed via an analysis of the *causes* of the situation. It is only by performing such social analyses that we can confront the reality. Such analyses quickly take you into many directions simultaneously: into anthropology, social psychology, economics, and the realm of how personal attitudes and social belief systems and values come about; into the political realm, too, of course, and into the global financial and political spheres. Such social analyses must lead to an identification of who has the power to make decisions (for example, on resource use), to control even the agendas of discussion, and to influence people's attitudes and behavior, be it through force or through manipulation.

Frequently, strategic analyses, especially by institutions dependent on public funding, will have great difficulty in indicating that there are those who benefit and those who suffer from maintaining the status quo, in the continuation of present systems of production, consumption, and distribution of goods and services. So, for example, you may end up with a conclusion of the need for greater "political will," as though this is an abstract phenomenon that either exists or does not, as opposed to accepting that such political will is a function of how economic and political forces operate and interact in a society.

Relationship with Nature

One of the things we have to develop is a fundamentally different attitude towards "nature." Regrettably, most societies that have adopted the industrial growth paradigm view nature in utilitarian terms, i.e., as a place from which to extract resources and as a garbage dump for wastes. No intrinsic value is placed on the conservation and preservation of plant or animal species and their habitats—on ecological diversity for its own sake.

A strongly hierarchical relationship is established between humans and nature—a relationship of domination and exploitation. Humans are seen as being separate from nature—not as inhabiting and forming a part of the web of life. Many hypotheses have been put forward as to why this is the case. Some blame the very foundations of Judeo-Christian thinking, which they say sets humankind apart from nature and gives our species instructions to dominate a harsh and ungiving nature—to extract a livelihood with the sweat of our brow. Others blame Cartesian thinking, which creates a duality between subject and object, between observer and observed. Yet others will say that the domination of nature arose because of social hierarchies which evolved in the earliest societies, even before the establishment of economic classes: hierarchies between young and old, male and female, rulers and ruled, and between those of different color.

Systems Thinking

So, to relate this to the sustainability of agriculture in Egypt: Reductionist scientific methodology has resulted in the creation of disciplines and specializations, and the subsequent creation of institutions which are highly segmented and disconnected. Biology, economics, engineering, sociology, politics, all come together to determine what happens in the hierarchy of agricultural systems—in the field, the farm household, the rural community, the nation, and the global commodity marketplace. Fortunately, efforts are now being made to pull different disciplines together—namely, multidisciplinarity and interdisciplinarity—but this falls short of developing a truly synthetic approach to understanding complex phenomena.

To promote *transdisciplinarity,* the whole education system will have to change. The difficulty, however, is that those who have to decide to change the educational structures are products of the system. These people must acknowledge that all is not well with the system that created them—then make a leap of faith, to make the difficult decisions needed to invent, test, and promote new approaches. This requires of them security in their sense of self to take these difficult positions. Yet it is absolutely necessary.

Social Goals and Economic Growth

Two other underpinnings of our societies have to be reexamined: those relating to economic growth and the role of the so-called free market.

Social development or social welfare has unfortunately come to be equated with economic growth. We talk incessantly about economic growth and development in the abstract. Does it make sense to count as economic growth those costs we incur in having to deal with the negative consequences of our production systems? Does it make sense not to factor into calculations of gross national product or national income the depletion of resources on which the income is based? Does all production contribute to improving the quality of life? Do all jobs really constitute good and useful work? The answer to all these questions in my view is: No! If we say that economic growth should not be our primary goal, then we need to be much clearer as to what our social goals actually should be. To answer questions about societal goals we have to look at our values. There is a close link between values and ethics; and between these and our economic and sociotechnical systems. We must try to be clear: What is valuable and who is valuable? So when we talk about development, we must ask: Development *of what?* Development *for whom?* as well as the more common, Development *how?* For further reading about alternative development models I recommend *Another Development* (Nerfin 1977).

Anybody who is serious about sustainability will ultimately have to confront these questions. We must realize that the situation we are in is not because of a deficiency in nature, or in our technical abilities. What are unsustainable are the demands that our value systems place on the environment, and on people. In our present system, the highest values are associated with the accumulation of material goods and the production and consumption of commodities. We need to ask again: How have these values become ingrained in us? But if accumulation is not to be our highest value, then somehow we have to become more balanced human beings, and find increased value and satisfaction in nonmaterial aspects of existence.

Bearing the Costs

Even a very superficial analysis shows that moving towards sustainable practices will involve significant costs. For those unwilling to bear the immediate costs, it should be emphasized that delays are likely only to increase those costs in the future. The problem is complicated by the fact that those who bear the costs and those who benefit may not necessarily be the same group: they could be separated both in time and in space. Costs, or benefits foregone now, may have to be incurred to benefit future generations; farmers upstream may have to bear costs to ensure benefits for those downstream; exporters may have to forego short-term profits; and everybody will be asking: Why me? In my view, the main sociopolitical question in our time will be how the costs for promoting conservation and enhancement of the natural resource base will be met, and who will bear them.

Redistribution: In the literature on sustainable agriculture, little mention is made of the need for a redistribution of wealth towards the agricultural sector, or towards the natural resource production and utilization system generally. Leaving aside subsidies that promote unsustainable production and the resulting dumping on world markets, how acceptable will such a redistribution be to those footing the bill? Clearly, a convincing case needs to be made of the importance of preserving the integrity of the natural resource base: But how? Enormous political forces keep prices of agricultural commodities low. To this must be added the complexity that the low prices in some way benefit poorer segments of society, since they spend more on food. It may be that it is becoming more important to ensure adequate income levels to the farming sector so that investments in sustainability can be made.

Military budgets as a source of funds: To ease the pain of the redistribution required to bring about sustainable economies, a concerted effort will have to be made to identify sources of surpluses for redistribution. The most obvious source of such surpluses globally is investments and expenditures made in arms production and the maintenance of military establishments, estimated at a trillion dollars annually. Some will say that when pressures on resource availability mount armaments will be increasingly needed. But therein lies the choice: the community of nations can make the a priori decision that peaceful resolution of conflicts and inequities will be sought. There is no real question as to which is the preferred choice. The shift to sustainability is desirable, as well as necessary.

The "Religion" of Free Markets

One of the reasons that the notion of "redistribution" receives so little attention is the quasi-religious attitude about the free market and, increasingly, "global competitiveness." To raise questions about the free market is almost to be accused of blasphemy, as Jeremy Seabrook (1991) indicates. Seabrook—a strong advocate of the Green movement in the United Kingdom—points out that the free market system, so much praised for its success in delivering "the goods," is separated from that which delivers "the evils." This creates a strange, distorted view of the world in which this particular economic system is somehow cleansed: it becomes innocent of all its impact on the world. The only flaw in the system becomes those faulty individuals "who stand in such shaming contrast to its shining perfection."

One of the main reasons for the near-religious fervor about market structures is the dismal failure of central planning systems over the past half century actually to deliver the goods. This view, too, is simplistic and superficial. It would be a terrible indictment of human imagination and creativity if we were really to be convinced that there is no road other than these two; no alternative modes of social and economic organization. Free market propaganda has got so out of hand in the euphoria and gloating over the demise of the so-called Communist regimes that for the moment it precludes any serious discussion of alternatives—and any really objective view of the impact of the market system.

Focusing on the market as the main mechanism for the exchange of goods and services has meant that what is not transacted in the market is considered not to have a value. All public goods are thereby devalued: clean air, clean water, functioning communities, social cohesion, lack of crime, and species diversity—these do not enter into the economic calculus. Unfortunately, economists have become the gurus of our age, and the worse the situation becomes (and the less they can deliver) the more our reliance on them seems to grow. We must get rid of the myth that traditional economics has all the answers to guide us out of the mess we are in. A new economics must be invented. Fortunately there are positive signs, however marginal, that this is happening. The writings of futurist Hazel Henderson (1978 and 1981) are a case in point.

The Sharing Mode

To return to the question of redistribution and how the prevalent economic dogma works against it: It has been drilled into us that in economic terms, if we all act to meet our own selfish objectives, the result will be not just individual welfare but social welfare as well. This, after all, is one of the bases of modern economic thought. It may have been a viable credo in the world of the past, an empty world very different from the crowded one we now live in, and a world where technologies did not have the power to inflict the damages that current ones can. But this view is not adaptive to the present reality (von Droste 1991). Therein lies the dilemma. After all these years of being brainwashed into believing that selfishness and competitiveness should be the bases of our behavior, we are confronted with the fact that without sharing, without balancing competitiveness with cooperation, without redistribution outside the market context, we will not achieve our goal of sustainability. Again new modes of interaction will have to be invented, tested, and implemented.

Economies which are built on the premise "grow or die" can only result in entropy and eventual suffocation. We are caught in a race for global competitiveness imposed on us by the very character of present economic structures. It is wonderful, in principle, to promote efficiencies in production. But if these technical efficiencies do not take into account social efficiencies and long-term environmental impact, the race for global competitiveness will inevitably lead to an acceleration of resource depletion and environmental degradation. It is highly regrettable that in the six years of the Uruguay Round of negotiations in the GATT there has been little consideration of the environmental impact of international trade policies.

Long-Range Outlook

These changes imply that humans become more consciously goal-seeking and purposeful. This in turn requires a long-range planning outlook, both temporal and geographic—something sorely lacking in modern societies and economies, accustomed as we are to the importance of "bottom lines." But it is clear that the horizon for decisions must change, from the short term to the long term, from the particular to general welfare. Following this line of thinking to its conclusion, we cannot but ask: Who makes the major decisions affecting society and local and global ecologies? Are our existing political structures and institutions adequate for long-term social planning? For what Don Michael (1978) calls future-responsive-societal-learning? Clearly not: So what do we need? This is a major question for debate and one which needs increased attention.

Recognition of Connectedness

Global economic and ecological interdependence and connectedness is a reality we must face. Individual choices made everywhere influence the global commons. Decisions taken by national or international bodies influence the conditions of individuals far beyond the obvious. Egypt, whatever it decides to do about the sustainability of its own agriculture, may be influenced by the decisions of billions of individuals around the world to continue to consume (or not consume) fossil fuels and emit greenhouse gases—an issue of particular interest to low-lying countries near the sea. If global warming does indeed lead to a rise in sea levels, the impact in Egypt would be great. Egyptians will also be influenced by those with whom it shares its watershed—be it an Ethiopian highlander, a Ugandan coffee grower, or the Government of Sudan. The cotton grower will be affected by speculation on the Chicago Commodity Exchange. The individual farmer in the most isolated rural setting will be affected by decisions taken in the executive bodies of the IMF, the U.S. State Department, and the European Economic Community. These distant decisions will influence resources available to the Egyptian government for investment, affect the prices of commodities, and so forth.

Interconnectedness extends to all spheres—economic, ecological, and ethical—and both space and time. We must therefore recognize that each activity an individual undertakes influences not only that individual's existence, but those of others, both human and nonhuman. This connectedness is not only in the material, physical plane: it extends to the intellectual, emotional, and spiritual dimensions of our existence. And it goes both ways.

In my view, being aware of these connections is a critical precondition for moving towards a more desirable, alternative future. I would like to assign a name to the transitional phase from the present mess to a more desirable future: The Age of Recognition of Connectedness.

Decision Time

In June 1992, the heads of most nations and thousands of other people met in Rio de Janeiro for the Earth Summit. It was a critical meeting and may have set the tone for years to come on how the human world will deal with the global problematique. Those who are trying to work out the important international conventions and frameworks for reduction in greenhouse gas emissions, maintaining biodiversity, and so on are likely to encounter many difficulties. I will not go into them here. They are well covered in a recent book (MacNeill et al. 1991).

MacNeill and his colleagues point out that one of the main characteristics of the situation is that many huge trade-offs will have to be made on an ongoing basis if we are to move in a positive direction. Deal-making will become a critical political capability. Going blindly down the path of "might is right" will lead to dead ends for all, both nationally and internationally. Notwithstanding the attractiveness of multilateral or global agreements, these authors also advocate reaching many smaller, "bilateral" bargains. I agree with this and would like to add that what is also much needed by some is the courage to take unilateral action, to break new ground and lead the way.

Trade-offs must be made not just between countries. International bankers and their investors must decide whether they want to continue their self-centered policies in resolving the debt crisis, when it means continued environmental degradation globally and ill effects, ultimately, for them as well. Nationally decisions will have to be made about what is needed; the costs of immediately promoting sustainable production practices must be undertaken. All countries will have to decide on whether the push towards global free trade and international competitiveness is indeed what they want, rather than more self-reliant, localized economies. Choices must be made between making short-term gains with major environmental costs and staying with long-term sustainable benefits. Decisions must be made globally on how to price cheaply extractable, but exhaustible or environmentally polluting, resources such as oil or coal; and on who pays for the costs of pollution. Choices will have to be made everywhere as to whether we want to maintain a vibrant, sustainable agricultural sector and rural economies, or continue to push people off the land and into cities because they cannot make a decent rural living. Much of this is premised on decisions about redistribution.

The Danger of the Authoritarian Scenario

The kinds of choices I have referred to are made primarily in the political arena. We must ask whether current political structures, systems, and institutions, both national and global, are capable of making the kinds of difficult choices to be made. Frankly, I do not think they are. And a major danger lurks here. The on-going environmental degradation will necessitate increased regulation of behavior if we are to avoid the ecological abyss. There is a great possibility that sustainability can be attained, but at the price of a major loss in personal and social liberties. I can imagine the development of enormous bureaucratic and technocratic structures to enforce "sustainable" behavior—behavior which should evolve through self-regulation. Murray Bookchin (1984 and 1989), the leading proponent of the "social ecology" movement, refers to this as the ecofascist or technofascist scenario.

It is important not to make the mistake of thinking that the ecological sustainability imperative has within its core the criterion of social justice. It does not. And that is why so many of those thinking about this issue insist that social justice be part of the discussion. The dangers are many. Huge inequities could be maintained by force and edict, by autocratic regimes in periods of economic decline, as Robert Heilbroner (1980) has pointed out. In many parts of the world, we have given up control over the decisions which intimately affect our lives to invisible bureaucrats, technocrats, and politicians. We have preferred to play the game of being active only in the accumulation of commodities while being passive in the political dimension of our existence. In fact we have been encouraged to do this by those who benefit from our nonparticipation.

Utopia or Dystopia?

Much of what I have proposed may appear to be utopian—dependent as it is on the emergence of a new type of human, and of new ways of interaction between humans, and between humans and nature. But this is what confronts us if we work through the internal logic of sustainability, through the implications of the goal of creating sustainable societies. Our choices are between the conscious creation of utopian societies on the one hand, and environmental disaster or authoritarian dystopias on the other.

References

Bookchin, Murray. *Toward an Ecological Society.* Montreal: Black Rose Books, 1984.
———. *Remaking Society: Pathways to a Green Future.* Montreal: Black Rose Books, 1989.

Conway, Gordon R., and Edward B. Barbier. *After the Green Revolution: Sustainable Agriculture for Development.* London: Earthscan Publications, 1990.

Heilbroner, Robert. *An Inquiry into the Human Prospect: Updated and Reconsidered for the 1980s.* New York: Norton, 1980.

Henderson, Hazel. *Creating Alternative Futures: The End of Economics.* Garden City, N.Y.: Anchor Press, 1978.

———. *The Politics of the Solar Age: Alternatives to Economics.* Garden City, N.Y.: Anchor Press, 1981.

MacNeill, Jim, Pieter Winsemius, and Taizo Yakushiji. *Beyond Interdependence: The Mashing of the World's Economy and the World's Ecology.* New York: Oxford University Press, 1991.

Michael, Donald. *On Learning to Plan and Planning to Learn.* San Francisco: Jossey-Bass, 1978.

Nerfin, Marc. Editor. *Another Development: Approaches and Strategies.* Upsalla: Dag Hammarskjold Foundation, 1977.

Seabrook, Jeremy. *The Myth of the Market: Promises and Illusions.* Montreal: Black Rose Books, 1991.

von Droste, Bernd. *Environmentally Sustainable Economic Development: Building on Brundtland.* Paris: UNESCO, 1991.

World Commission on Environment and Development. *Our Common Future.* Oxford: Oxford University Press, 1987.

19

Population Dynamics and Sustainable Agricultural Development in Egypt

Ismail Sirageldin

My purpose in this chapter is to answer the question: Do current policies and trends in Egyptian population dynamics and agricultural development represent a stable and sustainable system? The importance of the question follows from the premises that (1) there are key processes in population dynamics that must be integrated in the design of a strategy for sustainable agricultural development; (2) the relation between agricultural development and population dynamics cannot be elucidated in isolation from developments in other social and economic sectors of society, and in the case of Egypt, in isolation from developments in the labor markets in neighboring countries; and (3) these processes and relations have been conceptually underdeveloped and operationally underutilized.

The subject is wide and diverse. In its full dimension, it covers almost all aspects of social and economic development. My objective is limited: to bring into focus some critical and structural issues in this complex vortex that are relevant to policy and planning. No attempt is made to provide a comprehensive review of theoretical frameworks or an evaluation of the existing data. Rather, my purpose is to provide a systematic framework for such analyses and present a set of principles, conclusions, and policy options that can serve as a basis for policy debate.

Three Concepts

Three conceptual issues require careful examination: sustainability, agricultural transformation, and the role of population dynamics in a sustainable agricultural transformation.

Sustainability

The decades of the 1970s and 1980s brought about a major shift in the understanding of the potential strong micro and macro response to the

219

apparent food shortages (Timmer 1988: 319). However, it is the sustainability of such technical change, especially its diffusion in the developing countries, that is being challenged. Brown (1984) is pessimistic about the ability of technical progress to remove the "natural limits" on human activity. Timmer, on the other hand, puts his trust in technology and free markets to maintain and achieve sustainable development. These two alternative views of sustainability have been considered as being either "progressive" (Timmer), or "environmentalist" (Brown). However, there are agreements and disagreements among and within the two groups (Norgaard 1991).

The bottom line of the debate is how to resolve the desire for *intra*generational efficiency with that of *inter*generational equity in such use. In that context, sustainability becomes a matter of intergenerational equity. The conventional economic view on sustainability by focusing on internalizing externalities may not be adequate. It relies heavily on a market that may not provide the right signals. Such problems are more confounded in rural areas. Recent reviews of the functioning of rural labor markets indicate serious and pervasive cases of incomplete markets and market failures (Stiglitz 1985). They also indicate that progress towards an integrated conceptual framework that can deal with determination of wages and other contractual terms, labor mobility, and market failures has been hampered by the inability to integrate "all major interrelated rural markets—land, labor, credit—into a single coherent rural model" (Binswanger and Rosenzweig 1981: 54–55). Some economists, considering sustainability as a matter of intergenerational equity, propose imposing constraints on the use of resources and environmental systems by the current generation (Daly 1973).

Systems of Agricultural Transformation

Agricultural transformation should be viewed as an integral part of the interrelated process of structural transformation or change during the transition from a low-income rural economy to an industrial urban economy. Structural change by definition is a disruptive process. It is a creator and destroyer of values (Goulet 1992: 467–475). It implies that various segments of the economy grow at different rates. It also implies significant geographic, social, and economic mobility. Accordingly, some groups of population lose and others gain. During this process, policy actions and institutional changes are required to reduce the costs of the structural shifts and minimize the resistance to change. The role of the state in facilitating this transition can be instrumental, depending upon a complex set of economic and noneconomic factors.

Egypt's apparent failure to implement policy reforms has been variously explained. Tuma (1988: 1185–1198) blamed the slow economic development in Egypt on the presence of an institutional behavior pattern based on Indecision, Procrastination, and Indifference, dubbed as IPI. These were being bred, sustained, and perpetuated by the existing institutions of religion, family structure, land tenure, education, and government. It is not evident, however, from Tuma's analyses, whether the IPI pattern of behavior is exogenous or endogenous to the present stage of development. Richards (1991: 1721–1730) argues that Egypt's failure to implement policy reforms during the 1980s is a result of the government's skillful exploitation of its political role in the region and its ability to capitalize on that "strategic rent" by extracting favors from the United States and the IMF.

Agricultural transformation has historically been a process combining decline and growth. The experience of the now industrialized countries indicates that the share of agriculture in a country's labor force and total output declines as per capita incomes increase, while agricultural growth accelerates. This pattern was taken by some students of economic development as a requirement for economic growth (Lewis 1952). The transformation process seems to evolve through definable phases. Timmer (1988) specifies four such phases. In the first phase, as a result of significant investment in rural infrastructure, institutional changes, and new technologies, agricultural productivity per worker rises, creating a surplus. In the second phase, the surplus is transferred for the development of the nonagricultural sectors. The third phase starts the integration of the agricultural sector into the new industrial economy. In the final phase, agriculture becomes one of many sectors in the economy and loses its initial special status. Each phase indicates the pattern of resource flow from agriculture, namely financial and labor. Government intervention can greatly influence the size and timing of financial flows during the process of transformation. The size and timing of labor flows could be affected significantly by direct and indirect government policy, as well as by external and exogenous forces, e.g., international migration.

The general pattern of agricultural transformation is not debated. It is the dynamic nature of the process and the perceived pattern of its linkages to nonagricultural growth that does not receive consensus. In the stages-of-growth theory, the process of transformation is explained as a movement from a primarily agrarian to an industrial economy. Such movements are viewed as historically predetermined. In the dynamic, dual-economy models, jumps could occur. Peasants play a central role in the elimination of dualism through their "engineered" incorporation in the market. There are other frameworks that call for a large role of government in designing strategies and program implementation. They stress nutrition as a key objective for agricultural development and focus on maximum employment during the transformation process (Mellor and Johnston 1984).

It is evident that although the expected outcome of the agricultural transformation is similar in the various perspectives—a sizable growth in agricultural output combined with a decline in the share of agriculture in total output and employment—the social cost associated with each path (e.g., equity or unemployment) could differ significantly. That final outcome, however, implies in all cases a major structural transformation in the demographic behavior of the rural and urban populations: its natural increase, spatial distribution, skill formation, and occupational mobility. Failing to adjust in an orchestrated manner, the sustainability of agricultural transformation is questioned. Indeed, experience indicates that the nature and timing of the agricultural transformation process are conditioned by the demographic behavior of the populations.

The Role of Population Dynamics

It is evident, then, that the rural population must undergo significant adjustment during the agricultural transformation process. Specifically:

- The number of agricultural workers per hectare (L/A) is reduced from over 40 percent to below 15 percent of the total labor force
- The productivity of agricultural workers (Y/L) increases relative to that of land (Y/A)
- Rural fertility experiences a sustainable decline

The result of this sizable geographic and occupational mobility is a parallel flow of resources between the agricultural and nonagricultural sectors. The sectors are mutually supportive. The process, however, is not necessarily smooth. It cannot be accelerated beyond the capacity of the agricultural sector to develop; the capacity of the nonagricultural sectors to absorb the agricultural surplus labor; the mobility response of the labor force to acquire new skills and move to new locations; and the extent of the demographic response. This pattern reflects clearly the historical experience of agricultural transformation, for example, in the United States from 1880 to 1980.

I must stress that the path of agricultural transformation in the United States, and its implied role of population dynamics, depends on the structure of its resource endowment, technological development, and public policy. Clearly it is not the only pattern. As shown in Figure 19.1, six different possibilities exist, based on the interaction among population growth and density, available land, and the use of appropriate technology. In this, Hayami and Ruttan (1985) examine the relation between agricultural output per labor unit (labor productivity) and agricultural output per hectare of land (land productivity). The dynamic interaction of population growth and agricultural labor outmigration during the process of agricul-

Figure 19.1 The Role of Population Growth in Various Possibilities for Changing Land and Labor Productivities, 1960–1980

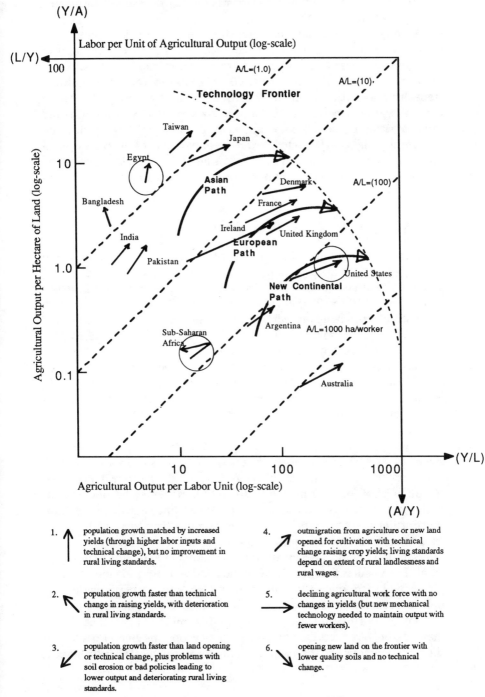

Source: Based on Timmer (1988) and Hayami and Ruttan (1985).

tural transformation is best illustrated by Timmer (1988).

Hayami and Ruttan identify three paths that reflect the international experience with agricultural transformation: (1) the New Continental, with the lowest labor intensity, (2) the European, with medium, and (3) the Asian, with the highest labor intensity. The technological frontier acts as an envelope that sets limits for these varied experiences. Attempts to reach beyond these limits implies a threat to sustainability. It is evident, however, that regardless of the path being followed, for the agricultural transformation to be completed, *structural* stability among population dynamics, growth in agricultural labor and land productivities, and growth in nonagricultural labor demand must exist. The emphasis is indeed on structural (long-term) stability. From the perspective of the discussion at the conference, it is an axiom that there is no agricultural transformation without population transformation, both in growth and distribution.

Egypt: Early Stages

My purpose is not to document the Egyptian experience with population and agricultural transformation, but rather to provide an outline for its general directions. Some basic trends in Egyptian agricultural development (from 1960 to 1980) are shown in Table 19.1. During that period, agricultural output increased from 17.7 million metric tons of wheat equivalent to 26.2 million. Agricultural workers (in terms of male workers) increased from 4.0 million to 5.6 million. Land under cultivation stabilized around 2.85 million hectares; livestock decreased slightly from 5.3 million to 4.7 million. There were serious attempts towards mechanization in Egyptian agriculture during the 1960–1980 period: tractor horsepower increased from 220,000 to 1,029,000. Fertilizer use more than doubled.

It is evident from these productivity trends that Egypt is in the early stages of its agricultural transformation. There has been a population response, but mainly in terms of geographic mobility. Movements of the Egyptian labor force during the 1970s and 1980s seem to be independent of the process of agricultural transformation outlined earlier in the chapter. Furthermore, government policies seem to have reacted to forces exogenous to the transformation process and to the achievement of an intersectoral balance, necessary for a sustainable agricultural development. The result is that in spite of all the good will and good intentions of policymakers to guide the country towards a sustainable growth path, policies seem not to have been coordinated to lead towards this objective. It is obviously a difficult task. We must first examine what are these forces that not only reduce the domain of policy options but also confuse short-term actions and solutions with long-term structural transformation.

Table 19.1 Selected Agricultural Statistics for Egypt, 1960–1980

Indicator/Variable	1960	1970	1980
Agricultural output 1000 wheat units	17,737	23,713	26,153
Number of male workers in agriculture 1000s	4,024	4,839	5,644
Agricultural land area 1000 hectares	2,569	2,843	2,848
Livestock 1000 livestock units	5,322	5,578	4,719
Fertilizer consumption $(N + P_2O_3 + K_2O)$ 1000 metric tons	204	331	526
Tractor horsepower 1000 hp	220	645	1,029
Number of farms 1000s	2,946	3,285	3,432
Literacy ratios percent	25.8	33.5	44.0
School enrollment ratios percent	36.0	50.0	59.0

Source: World Bank, *World Development Report* (various issues) and Hayami and Ruttan, 1985.

They are related to: (1) population, (2) agricultural policy, and (3) intersectoral relationships.

Population-Related Factors

The main focus in examining the role of population in agricultural transformation is its geographic response and its growth relative to vertical and horizontal expansion of agricultural production. Trends of population growth in Egypt raised a puzzle during the 1970s and 1980s. The crude birth rates, after a significant decline in the 1960s, started an increase in the early 1970s. This was confused with change in fertility. A careful analysis by Bucht and El-Badry (1986) has shown the hazard involved in relying on the crude birth rate as an indicator of fertility trends. They have provided convincing evidence that "both the decline of the birth rate

during the 1960s and early 1970s and the subsequent increase were substantially influenced by structural factors, namely earlier changes in fertility and mortality." After adjusting for these structural factors, they have
shown a slow but systematic decline in Egyptian fertility. The decline almost
stagnated by the late 1970s and early 1980s. Also, there are significant
rural-urban differentials. The growth momentum is sizable and will persist
for a long time. Furthermore, recent analyses indicate that by the mid-1980s,
Egypt was not perceived to have entered the sustained phase of a declining
fertility indicated by the demographic transition (Zaky, Wong, and Sirageldin
1991).

The figures showing a lack of significant and sustained decline in fertility
raise serious concerns as to the country's ability to achieve sustainable
agricultural transformation in the near future. It is the population movement,
however, that is puzzling and potentially confusing in the Egyptian experience. This movement, however, has not been part of the equilibrating dynamics expected in the process of sustainable agricultural transformation.
Farmers and agricultural workers were not responding to signals from opportunities generated in growing domestic, nonagricultural sectors. Most of the
response has been to outside signals, mainly from the Gulf countries. The
response has been large but temporary. It should not be confused with the
process of outmigration discussed above. In this latter process, migrants are
not expected to return to their initial occupation or original place of residence.
They are part of a passing phase. This evidently has not been the case in Egypt.
The basic structural characteristic of this temporary outmigration is well
documented by Fergany (1992) and Saadani (1991).

It is evident from these studies that the incentive to migrate has been
high. Almost half the low-income migrants could earn fifteen times their
original incomes (wages) in Egypt. Migrating women, although relatively a
small proportion of those involved, start out with a smaller average income
but earn almost fourteen times as much at destination. Women also had more
than double the years of education of migrating males. The implication for
the household roles and decisions needs more investigation. Agricultural
workers comprised 37 percent of the total migrants examined. Their income
at destination was almost ten times as much as nonmigrating agricultural
workers. The majority of migrating agricultural workers are illiterate, and
many of them are recent migrants.

This type of migrant and the dynamics of the migration are clearly
different from the expected pattern in an orderly agricultural transformation
process. Instead of being the type that would necessarily lead to the stability
of the transformation process, it has the potential to provide signals that may
lead to policy initiatives in conflict with this aim. For example, during the
1980s the government essentially treated international migration as a major
source of foreign exchange and developed policies to maximize remittances.
At the same time, its population policy was weak, operationally and conceptually.

Agricultural Policy

Many studies deal with agricultural policy in Egypt (e.g., Dethier 1989). My purpose is to examine the expected role of the population factor as implied in the stated policy objectives. Dethier lists seven policy objectives, the weights attached to them, and their expected effect on producer prices. None of the objectives deals explicitly with the role of population dynamics in the transformation process and none of the expected effects deals with the consequences to population movement or growth. Actually, Dethier finds that policies supported the landless; heavily taxed the real income of farm households; and promoted a sizable transfer of agricultural income. Examined as an agricultural-demographic interacting system, these policies would not necessarily lead to the sustainability of the transformation process.

Intersectoral Issues

During the past two decades, Egypt has experienced sizable external shocks, some positive and some negative, that have affected policy priorities. The country is now a net exporter of oil. Revenue from oil, when combined with that of remittances and foreign aid, has become a major source of foreign exchange. The share of agriculture in GDP declined from 29 percent in 1965 to 19 percent in 1989 (World Bank 1991). This apparent decline in the share of agriculture in the total value added should not be confused as an outcome of a sustained process of agricultural transformation, in which the absolute real value added of agriculture increases substantially while its share in the total value added and employment declines.

The present dynamics seem to change the role of agriculture in the economy (as envisaged in the early 1980s) as the engine for economic growth in the country (Wally 1982). By the late 1980s, and especially after the Gulf war, it seems to have taken a rather passive role. "Dutch Disease," well known as a resource misallocation problem, may well be operating in Egypt under the combined effect of oil revenue, remittances from temporary migration, and foreign aid. The situation is complicated by mounting public debt and an increased reliance on foreign aid, mainly from the United States. In 1990, the ratio of Egyptian total debt outstanding and disbursed as a share of GNP was about 127 percent (World Bank 1992: 60).

The Total Vision

In this chapter it has been argued that sustainable agricultural transformation is an integral part of the overall process of sustainable development. It includes, as a necessary condition, a full transformation of the demographic system in terms of growth and spatial distribution. These changes are structural and long-term in nature. Policymakers should have a vision of the total process in order not to confuse short-term fluctuations with long-term objectives. Egypt is in the early stages of its agricultural and demographic transformations. It faces, however, sizable external shocks that confuse the ability of the national market to allocate its labor and nonlabor resources spatially and temporally.

These conditions create a difficult environment for policymakers to plan for the long term, in which sustainable agriculture is a major concern. Its interaction with the process of population dynamics is structural. The problem is that in Egypt the short-term issues (crises) have taken precedence. The success of the policies adopted is a matter of survival. Indeed, all governments are concerned about balancing political stability, equity, and efficient utilization of resources. However, sustainability is an issue of intergenerational equity, and attempting to buy political stability today at the expense of instability tomorrow is not a sustainable strategy.

An important conclusion of the analysis is that, once a vision of the total process is formed, it would be possible to develop a concerted set of policies that enhance the transformation process without compromising sustainability. Policies should be directed towards transforming labor migration into labor-intensive export industries that must be combined with a vigorous demographic transformation policy. For example, public incentives may be shifted from individual rewards (aimed at attracting migrants' remittances) to those that encourage national enterprises to export, initially, labor-intensive goods and services, and, ultimately, high value-added products. A vigorous population policy is clearly essential for the long-run success of such an integrated policy package.

Furthermore, the presence of an efficient market is clearly necessary for a sustainable transformation process. We must, however, realize that there are limitations to relying fully on market solutions.

References

Binswanger, H.P., and M.R. Rosenzweig. *Contractual Arrangements, Employment and Wages in Rural Labor Markets: A Critical Review.* New York: Agricultural Development Council, 1981.

Brown, L.R. "Overview," in *State of the World, 1984.* Edited by L.R. Brown et al. New York: Norton, 1984.

Bucht, Birgitta, and M.A. El-Badry. "Reflections on Recent Levels and Trends of Fertility and Mortality in Egypt," *Population Studies,* 1986, Vol. 40, No. 1, 101–113.

Daly, Herman E. Editor. *Toward a Steady-State Economy.* San Francisco: W.H. Freeman, 1973.

Dethier, Jean-Jacques. *Trade, Exchange Rate, and Agricultural Pricing Policies in Egypt.* Vol. I, The Country Study. World Bank Comparative Studies. The

Political Economy of Agricultural Pricing Policy. Washington, D.C., 1989.

Fergany, Nader. "Arab Labor Migration and the Gulf Crisis." Cairo: Draft mimeo, 1992.

Goulet, Denis. "Development: Creator and Destroyer of Values," *World Development*, 1992, Vol. 20, No. 3, 467–475.

Hayami, Y., and V.W. Ruttan. *Agricultural Development*. Baltimore: Johns Hopkins University Press, 1985.

Kuznets, S. *Economic Growth of Nations: Total Output and Production Structure.* Cambridge, Mass.: Harvard University Press, 1971.

Lewis, W.A. "Economic Development With Unlimited Supplies of Labor," *Manchester School of Economic and Social Studies*, 1952, Vol. 22, 139–191.

Mellor, J.W., and B.F. Johnston. "The World Food Equation: Interrelations Among Development, Employment and Food Consumption," *Journal of Economic Literature*, 1984, Vol. 22, 531–574.

Norgaard, Richard B. *Sustainability as Intergenerational Equity: The Challenge to Economic Thought and Practice.* Internal discussion paper, Asia Regional Series, Report No. IDP 97, Washington, D.C.: World Bank, 1991.

Richards, Allan. "The Political Economy of Dilatory Reform: Egypt in the 1980s," *World Development*, 1991, Vol. 19, No. 12, 1721–1730.

Saadani, Somaya. "On the Determinants of Egyptian International Migration." Seminar paper, Department of Population Dynamics, Johns Hopkins University, Baltimore: 1991.

Sirageldin, Ismail. "The Potential for Economic-Demographic Development: Whither Theory?" *Pakistan Development Review*, 1986, Vol. 25, No. 1, 1–42.

Stiglitz, Joseph E. "Economics of Information and the Theory of Economic Development," NBER working paper, Washington, D.C., February 1985.

Syrquin, Moshe. "Patterns of Structural Change," in *Handbook of Development Economics*. Edited by Hollis Chenery and T.N. Srinivasan, Vol. I. Amsterdam: Elsevier Science Publishers, 1988.

Timmer, C. Peter. "The Agricultural Transformation," in *Handbook of Development Economics*. Edited by Hollis Chenery and T.N. Srinivasan, Vol. I. Amsterdam: Elsevier Science Publishers, 1988.

Tuma, Elias H. "Institutional Obstacles to Development: The Case of Egypt," *World Development*, 1988, Vol. 16, No. 10, 1185–1198.

Wally, Youssef. "Strategy of Agricultural Development in the Eighties." Paper submitted to the Conference on the State of the Economy. Cairo: 1982.

World Bank. *World Development Report 1991*. New York: Oxford University Press, 1991.

———. *Global Economic Prospects and the Developing Countries*. Washington, D.C.: World Bank, 1992.

Zaky, Hassan, Rebeca Wong, and Ismail Sirageldin. "Testing for the Onset of Fertility Decline: An Illustration with the Case of Egypt." Discussion paper, Department of Population Dynamics, Johns Hopkins University, Baltimore: 1991.

20

Integration of Environmental and Sustainable Development Dimensions in Agricultural Planning and Policy Analysis

Fahmy K. Bishay

A major challenge to the populations of developing countries is to alleviate poverty and create material abundance as has been achieved in the more prosperous areas. Since most of these developing countries are predominantly agricultural, sustainable development of their agricultural sectors becomes essential to their economic progress. The planning of sustainable agricultural development is a process of maximizing the sector's contribution to the economic welfare of the society, taking into account its special features, its interrelationship with other sectors of the economy, as well as the conservation of its natural resource base for sustainable use by future generations. This process involves various types of problems and controversial issues that render sustainable agricultural development a complex process of a multidimensional structure. Some of these problems and issues will be briefly mentioned here.

The nature of the sector makes it less amenable to development programs than most nonagricultural sectors. Agricultural production is highly dependent upon soil type, weather conditions, and various natural factors. Long "gestation lags" between investment decisions and the actual addition to productive capital stock are a characteristic of many agricultural projects (e.g., in Egypt land reclamation programs, big irrigation schemes, trees, forestry production, and livestock breeding). Such investments are especially vulnerable to weather and other natural conditions during their maturing periods, which raises the risk to high levels. Production follows a seasonal pattern and farmers are forced to be idle during a part of the year; they must also be skilled in a variety of seasonal tasks. Further, the production structure dependent on large numbers of individual farmers complicates drastically all attempts at promoting sectorwide economic and social improvements. The noneconomic variables operating in the sector represent a crucial and complex feature in sustainable agricultural development. The type and size of holdings and land-use involve political decisions. The realization of surplus production requires understanding of the sociocultural values of the farmers.

The nature and strength of the interrelationships of the agricultural sector with the rest of the economy change as development progresses. The problem of industrial versus agricultural development has been seen as a dilemma in the literature on economic development. Industrialization must be supported by adequate food supplies. Unless food is imported—which creates another problem—domestic farm outputs have to be increased, but this requires resources that then cannot be used for industrial development. The increase in agricultural production depends upon several factors, such as its share in national investment, allocation of this investment to alternative activities, and technology used in farm production and its organization and management. Technology and management are intimately related. The success of modern agricultural techniques is a matter not only of the willingness of farmers to use them, and their skill in implementing them, it also depends upon such organizational elements as adequate supply of production credit and development of efficient marketing systems.

The spatial aspect is probably a more important issue in agricultural development than in any other sector of the economy, due inter alia to the overwhelming role of land. The natural resource base varies greatly in different regions, especially in large countries. As a result, the same agricultural product can have different production functions in different regions. The determination of the optimum agricultural product mix in different regions, therefore, is of prime importance in sustainable agricultural development. Another reason underlining the need to take into account the element of space in sustainable agricultural development derives from the differences which sometimes exist among regional interests. Different regions may have varying levels of development and different potential. For the realization of a more equitable regional income distribution, the determination of agricultural investment and production plans should explicitly account for the regional differences at various levels of development.

Planning Agriculture in a National Framework

In the past, the agricultural sector was often viewed as a passive partner in the development process. It was widely believed that industrialization was the vital strategy for development, and that the leading manufacturing sectors would pull along with them the backward agricultural sector. Industry would provide a source of alternative employment for the rural population, generate a growing demand for food and agricultural production, and supply the agricultural sector with the industrial inputs (e.g., machinery and fertilizers). Accordingly, the bulk of investment in developing economies was directed to the industrial sector, while agriculture was assigned the passive function of being a potential source of unlimited labor and agricultural surplus for the rest of the economy. As a consequence, a popular policy prescription to encourage this strategy of transferring labor and agricultural surplus was to turn the terms of trade against agriculture. But, as indicated by Thorbecke (1969), "the trouble with this approach was that the backward agricultural goose would be starved before it would lay the golden egg."

Recently it has become evident that the functions that the industrial and agricultural sectors must perform in the process of economic growth are interdependent. That is, agricultural development planning must be fully integrated in any comprehensive multisectoral approach to the planning of national development. Thus governments of many developing countries have assumed major responsibilities for setting up and implementing socioeconomic development programs.

Sustainable Agricultural Development Objectives

National development objectives and objectives of sustainable agricultural development are interdependent with mutual feedbacks. The following are the main objectives of sustainable agricultural development.

Contribution to GDP: A major objective of the agricultural sector in most developing countries is to maximize its contribution to the gross domestic product (GDP). This need derives from the national development objective of achieving a sustained annual growth rate of GDP. The performance of the agricultural sector is mainly judged by its ability to expand the production of food and raw materials. In developing countries, increasing agricultural production is of vital importance if the growing demand due to growth of both population and income is to be met. This is particularly true if the poorest segments of the population share in the income growth, since they have a high income elasticity of demand for food. Rising income also generates the demand for a diet of higher quality and more variation, so that the pattern of agricultural production would need to be adjusted.

Population and employment: Most developing countries face a pressing need to provide employment for their rapidly growing population. Egypt is no exception. Overpopulation has, among other effects, a negative impact on the quality of the natural resource base. In agriculture, underemployment becomes a serious problem. Nevertheless, the task of creating employment is not exclusive to nonagricultural sectors. With the rapidly growing population and the inability of the nonagricultural sectors to absorb the unemployed labor, the agricultural sector cannot be disregarded in the national employment objective. Different types of unemployment as well as various stages of economic development imply different employment strategies on both national and sector levels. Hence, a determination of the optimum agricultural employment strategy should be specified and incorporated into sustainable agricultural development programs.

Income distribution: Since poverty and overpopulation encourage over-use and depletion of natural resources, a major objective of sustainable agricultural development is the rapid reduction in rural poverty and improved income distribution. Sustainable agricultural development, therefore, calls for reducing the variance in income distribution both spatially (regional) and temporally (intergenerational). An important aspect of the problem of wide income disparities, especially in developing countries, is the remarkable difference between rural and urban incomes. The relative progress of the agricultural sector, therefore, has a direct effect on improving the income distribution patterns in these countries.

A high degree of income disparity can be also observed within the agricultural sector. Among the rural population, there are large numbers of subsistence farmers, small farmers, and laborers (and their families) whose income is much lower than that of the relatively few landlords and large commercial farmers. Their poverty reflects the fact that their productivity is low and that they own few assets in the form of land, capital equipment, or human capital. Income disparity can also be observed between the different regions of the agricultural sector. Concerning poverty alleviation, the main antipoverty strategies include: public works; food distribution; and creating or distributing assets to the poor. The key to the success of these measures is the effective targeting at the poor. These options also affect environmental stability.

Balance of payments: The agricultural sector has an important role in the foreign trade policies of most developing countries. A majority of these countries are still largely dependent on primary exports as their major source of foreign exchange. Their import capacity is, therefore, significantly influenced by the flow of these exports. Although exports of minerals and manufactured goods have grown more rapidly over the last decade, agricultural primary products still account for a large proportion. The dependence of many developing countries on their agricultural exports creates several problems for sustainable development. The unit value of such exports from both developed and developing countries has almost stagnated over the last decade, but in terms of volume, agricultural exports from developing countries have grown far more slowly than those from the developed areas. As a consequence, the total value of developing countries' agricultural exports grew at a much slower rate than that of developed countries. It follows that the terms of trade of the developing vis-à-vis developed countries are not improving.

Environmental protection and resource conservation: By its nature, the agricultural sector depends upon land, water, other natural resources, and human resources for its biological production system. It follows that the sector plays a major role in the conservation or depletion of these natural resources. As stated in a document of the Netherlands Conference on Agriculture and the Environment (FAO 1991: 8):

> A policy goal for sustainable agriculture should be to provide or enhance incentives for farmers and rural dwellers to conserve and increase the natural capital stock, to maintain basic natural cycles (water, carbon, nutrients), and to contribute to providing renewable energy supplies.

The same document further indicates that the agricultural sector also plays a major role in the protection or degradation of the environment as a whole. Therefore, it is recommended in the document that protecting the environment should be included among the sector's major objectives. More specifically, the document suggests that

> the capacity of agriculture to recycle wastes and effluent of human activities must also be fully used by returning to the soil in some appropriate form the soil nutrients and organic material which are exported to industry and urban areas as food and other agricultural supplies.

The Planning-in-Stages Approach

In the macro stage of planning, a macroeconomic study of the behavior of the main strategic variables affecting the general performance of the entire economy is carried out, using an appropriate macroeconomic model, e.g., the Harrod-Domar model. The instruments and the extent to which they are to be used to achieve the main aims of development are broadly decided upon. Among the most important objectives of almost all types of economic policy are the increase in national income or GDP; higher employment level; more equitable income distribution; and a reduction in the balance of payments deficit. The national objective of environmental protection and sustainable development should be explicitly introduced in this stage. The most important means considered are investment—roughly including material investment (capital) and human investment (education and training)—and taxes. Major instruments for environmental protection and sustainable development should be explicitly considered, as is shown later in the chapter.

At the macro stage, the national development strategy of the country is developed. Environmental and sustainable development considerations should assume high priority in the strategy. To this end, the general principles—based on the discussions at the 1990 Bergen Conference on

Action for a Common Future—should be carefully taken into consideration. For the implementation of these principles, the basic concepts of national income accounts have to be revised. In this regard, many serious attempts have already been started with a view to developing an environmental accounting system. The following is a brief discussion of the key issues involved and the relevance to sustainable agricultural development planning and policies.

A fundamental step that governments could take in promoting more sustainable development is to revise current national income accounting systems to fully reflect the importance of natural resources as economic assets. Other environmental indicators are also needed to measure changes in environmental quality as well as the costs of environmental protection. While some costs can be monetized and linked to the national income accounting framework, other indicators, such as those related to biological diversity, can more appropriately be kept in physical terms. The main objective of a national environmental accounting system is to correct the bias in the economic indicators of environmental scarcity. This bias is mainly due to the existence of externalities, the structure of property rights, limits to information, and the frequently observed bias in pricing signals as a result of government intervention in key product and factor markets. Thus, reliance on such biased market indicators would lead to a fundamental discrepancy between "privately rational" and "socially rational" use of environmental resources.

The Middle Stage of Planning

In the middle stage of planning, the picture resulting from the macro stage is made clearer as the assumption of homogeneity of production and nondistinguishability among different sectors is dropped. The most important problem in this stage is to determine the appropriate growth rates required for the sectors in order to achieve the target of the national sustainable growth rate ascertained in the macro stage. The number of sectors dealt with in this stage should be reasonable; and they must be as homogeneous as possible. The defined sectors are usually assumed to be related to each other according to the traditional Leontief input-output structure. The sectors of the economy may be classified into international and national sectors. The products of the international sectors (international products) are tradable abroad, while those of the national sectors (national products) are not internationally tradable.

In the middle stage, three main uses of the input-output approach for the analysis of the agricultural sector are usually carried out.

1. Analysis of the economic structure of the sector within the national economy. Structural analysis aims at studying the properties of a

given model in a particular structure.
2. Formulation of economic policies, or the program of actions. This requires an analysis of the effects of a given type of action on certain economic variables.
3. Prediction.

The main distinction between policy formulation and prediction is that for the latter some analyses must be made of all factors affecting a given outcome. Policy formulation may, therefore, be conceived of as a conditional prediction. The two become identical only in the event that the conditions assumed in the policy prescription are fulfilled.

The Micro Stage of Planning

In the micro stage, or project phase, a detailed subdivision of sectors into various projects is carried out. Thus, the analysis goes into still more detail as it deals with projects, rather than aggregate sectors. The main objective in this stage is to select the projects that fit into the pattern decided upon in the middle stage. Since projects are much more specific and disaggregated than sectors, data available for their analysis can also be expected to be more accurate and more detailed. This may even lead to improvements in the coefficients utilized in the macro and/or middle stages.

In actual planning, a number of important factors have to be taken into account. The following are some examples:

- Projects in progress must have special priority
- There should be room for projects which have not yet been defined
- Project data must cover the gestation, learning, and operation periods
- Other social, environmental, and sustainability targets should also be considered, e.g., health; social security; natural resource conservation; and environmental protection

Agricultural Policy Analysis

Policymakers are often unaware of the impact of agricultural policies on soil erosion, soil nutrient mining, forest destruction, and marine pollution; or of the national income and foreign exchange savings that could be generated through the adoption of measures for sustainable agricultural development. In many countries, therefore, one observes what may be called "policy failure," in the sense that certain policies, which are not directly designed to influence the natural resource base, have had negative impact on the conservation and maintenance of natural resources. There are many examples of policy failure: in some countries subsidies have resulted in excessive use of chemical fertilizers, with negative impact on soil conservation; heavy subsidies have resulted in overuse of irrigation water, leading in some cases to waterlogging and increased soil salinity.

The Missing Factor in Price Policies

Agricultural price policy is seen by most governments as a vital means of affecting agricultural output and income, consumer welfare, and, by affecting the cost of living, inflation, employment, and growth in the economy. It is used as a major source of government revenue in most developing countries. Perhaps this factor is the main reason why governments intervene in agricultural prices and trade, creating domestic price levels which differ from internationally determined commodity prices.

The objectives and goals stated by most countries for agricultural price policy are basically similar. The objective of enhancing environmental protection and sustainable development is almost always missing. This objective is extremely important and should figure prominently in agricultural price policies. Measures to achieve this objective are discussed below.

Economic policy instruments can dramatically lower the costs of achieving environmental goals by allowing producers and consumers to decide how best to meet them at least cost. More importantly, broad, market-based incentives can bring about the adjustments needed to deal with complex, long-lived, dispersed environmental problems associated with various consumption and production patterns. The long-run responses to such incentives are much greater than the short-term effects. Rapid, environmentally benign technological improvement, which is essential if living standards for all are to improve, can best be stimulated through market-based economic incentives, especially price policies. Rational pricing of resources and resource-based commodities, reflecting full marginal supply costs, including costs of environmental protection, is the appropriate starting point for the use of economic policy instruments.

Governments use a number of policy measures in affecting prices, but two of them are evidently the most commonly used—often with great impact. The first focuses on producer prices. While few countries utilize multiple criteria in setting fixed prices, in most countries cost of production is the basic criterion used. In Egypt, producer prices for very few crops are presently determined by the government (notably cotton and sugarcane). In these cases, official prices are based on multiple criteria with particular emphasis on production cost estimates by crop in each governorate, taking into account the cost of labor and land, and adding an adequate profit margin. Producer price incentives could also be utilized to support those crops whose production is least damaging to the environment or "exhausting" to the natural resource base; e.g., tree crops and legumes tend to have more positive effects on the environment than do annual crops such as cotton.

The second policy measure is in dealing with input prices. Input subsidies are widespread in almost all developing countries. They are usually justified on the basis that they provide incentives for the adoption of new technologies, the use of associated modern inputs, and compensation for low producer prices. Although input subsidy schemes are established to enhance productivity and compensate farmers for low, fixed commodity prices, in general they have not accomplished the latter and in many countries the result has been a transfer of income from agricultural to nonagricultural sectors.

In Egypt, an analysis was made of the new burden of price policies on the agricultural sector in 1980. The exercise revealed that although input subsidies were relatively large, approaching £E 407 million, the producer price transfer was even larger, amounting to £E 987 million. The net burden on the agricultural sector was about £E 580 million and the input subsidies compensated for only 41 percent of the "tax" imposed on agriculture through price policy. These relations have, however, changed in Egypt because of price reforms initiated since the mid-1980s.

It is particularly necessary that the need for reconciliation among multiple price policy objectives and trade-offs be recognized. Experience shows that an initial failure to effectively formulate policy in this regard only too often leads to expectations for sustainable production and consumption which are not met, as well as to serious aggravation of budgetary and foreign exchange problems. However, most countries do not appear to take all of these factors into consideration; nor do they adequately investigate the impacts, costs, and benefits of alternative approaches in the formulation of their agricultural price policies for sustainable agricultural development.

In addition to the need for continuity and credibility in support of price policy implementation, other factors include: (1) institutional coordination, (2) manpower resources, and (3) information and data.

Environmental Protection and Public Expenditure

With the structural adjustment program being implemented in many developing countries, including Egypt, the rationalization of public expenditure assumes high priority. However, it is especially advisable to avoid indiscriminate cuts in various items of public expenditures, with particular attention placed on enhancing (rather than reducing) investment expenditure. To this end, four suggestions can be made—based on the discussions at the 1990 Bergen Conference on Action for a Common Future (Bergen 1990)—for the integration of environmental and sustainable development considerations into public expenditure at the macro stage.

1. Public spending on research and infrastructure markedly influences technological development. Governments should consider early in the planning process the long-term sustainability of technological systems supported by public spending.
2. Public expenditures on natural resources are strongly biased toward expanding supplies of marketable commodities, often at the expense of the nonmarketed services those resources provide.
3. The costs of preventive measures are usually much lower than the costs of repairing or coping with the environmental damage/natural resource disaster. Thus, governments should ensure against possible catastrophic and irreversible ecological disruptions.
4. To ensure that public investments effectively lead to sustainable development, people's participation in the planning process, through enhanced decentralized regional planning, is essential.

International Trade Policies and Distortion

International trade policies have contributed significantly to a highly distorted world agriculture. In fact, agricultural trade policies in both developed and developing countries have jointly resulted in an international pattern of trade distortions between trading partners the world over. The developed countries generally provide a high level of protection and subsidies to their agricultural sectors. The developing countries, on the other hand, generally do the opposite. They often discriminate against their agricultural sectors by shifting their domestic terms of trade against agriculture. The main instruments used by developing countries in this regard include overvaluing their national currencies, imposing export taxes, and setting up price ceilings on strategic food products.

These policies encourage a substantial share of the world agricultural output to be produced in the "high cost" developed countries, and a very small share in the "low cost" developing countries. This inefficient use of the world agricultural resources could further lead to environmental damage in both groups of countries. The relatively high prices in the developed countries tend to encourage an "overuse" of fertilizers, pesticides, water, and other inputs. This, in turn, leads to the pollution of underground water, the pumping of underground aquifers, and the bringing into production of marginal lands. In the developing countries, overvalued currencies—which act as implicit import subsidies—tend to encourage the developed countries to continue their distortionary policies, since they reduce the cost of those policies. Meanwhile, overvalued currencies acting as implicit export taxes, combined with other policies, encourage a pattern of agricultural production that is more extensive, especially for small farmers, than would otherwise be, and could, in turn, induce natural resource degradation and environmental damage.

Institutional Reform for Sustainable Development

Effective natural resource management for sustainable agricultural development calls for regional decentralization of agricultural planning and policies. An FAO study (1988) on the subject identified many advantages of institutional reform, emphasizing decentralization. To assist governments in accelerating the process of regional decentralization for sustainable agricultural development planning, the same FAO document established a set of operational guidelines to achieve this objective.

In addition to decentralization, the following general principles for strengthening institutions—discussed at the 1990 Bergen Conference on Action for a Common Future—are also useful. These principles imply that:

> Many natural resources are managed more efficiently when users have secure property rights. Open-access common property resources are typically exploited excessively and inefficiently. Governments should examine the extent to which individual and community rights over natural resources can be established or strengthened in ways that lead to greater equity, efficiency, and sustainability in their use. . . . Countries expanding the scope of the market economy must establish a system of *laws and environmental regulations* and evenhandedly enforce them as a basic precondition for the functioning of the economy. Moreover, in order to allow market incentives to operate effectively, governments must establish rules that allow enterprises to retain efficiency gains and that make them responsible for losses, rather than providing budgetary or financial subsidies.

References

Bergen Conference on Action for a Common Future. Report of the Conference. Bergen, Norway: 1990.

FAO. "Issues and Perspectives in Sustainable Agriculture and Rural Development." Netherlands Conference on Agriculture and the Environment, Main Document 1. Rome: FAO, 1991.

———. *Regional Decentralization for Agricultural Development Planning in the Near East and North Africa.* FAO Economic and Social Development Paper 73. Rome: FAO, 1988.

Thorbecke, E. *The Role of Agriculture in Economic Development.* New York: Columbia University Press, 1969.

21

Food, Jobs, and Water: Participation and Governance for a Sustainable Agriculture in Egypt

Alan Richards

The sustainability of Egyptian agriculture depends on the sustainability of wider economic developments. Agriculture is now a minority sector of the Egyptian economy, contributing less than a fifth of national income and exports earnings and employing one-third of the labor force. Agriculture is located within an economy which is undergoing a painful yet necessary period of restructuring. Because of the essentially supportive, rather than leading, role of agriculture in the national economy, the sustainability of Egyptian agriculture depends on the success of the structural adjustment process underway.

Egypt has been living beyond its means for a long time.[1] For nearly twenty years, it invested more than it saved, and bought from abroad more than it sold. The government spent more than it received. These macroimbalances were exacerbated by pervasive microdistortions, which thwarted efficient resource allocation by sending socially erroneous signals to producers and consumers. The regulatory environment blunted incentives, particularly in the private sector. Resulting patterns of production were very far from Egypt's comparative advantage. The savings, trade, and fiscal gaps posed an increasingly serious threat to the sustainability of the national economy as the 1980s drew to a close. In the final analysis, the gaps were plugged by borrowing from foreign governments: international debt rose from perhaps US$2 billion in 1970, to US$21 billion in 1980, to just under US$50 billion on the eve of the Gulf crisis in 1990. What was at first a benefit became a cost as debt service consumed ever larger amounts of scarce foreign exchange.

There are, however, deeper reasons why structural adjustment is essential to the sustainability of the Egyptian economy. Economic reform offers the only hope of solving the problems of job creation and poverty alleviation. During the 1990s, about six million young people will enter the labor market. If the demand for labor does not grow along with this supply, either unemployment will rise or real wages will fall, or (most likely) some combination of each will happen. In any case, poverty will increase. Yet during the past two decades, incentive structures inhibited job creation. In fact, perhaps 90 percent of all jobs created from 1976 to 1986 came from the government and international migration, two sources which could not (and cannot) be maintained.

Egypt will also have to become a more efficient participant in the international economy simply in order to buy food. Rising water scarcity (which is now probably a more important constraint on agricultural output than land scarcity) means that Egypt will have to rely heavily on international trade for food security. This, of course, does not mean that agricultural output cannot or will not increase, nor that domestic food production cannot play a leading role in the "food security portfolio" of the country. But it does mean that the goal will have to be food security and not self-sufficiency. Egypt cannot afford to try to be self-sufficient in all foods. Food security (which increasingly means foreign exchange security) and job creation must be the centerpieces of any strategy for a sustainable national economy.

Agricultural development can support this wider process of transition to sustainability in four main ways. First, agriculture could make a significantly greater contribution to the generation of foreign exchange than is now the case, through the enhanced production of export and import-substituting crops. Foreign exchange will also be saved through more efficient use of inputs. Second, the sustainability of an economically efficient agriculture coping with an expanding population and (hopefully) rising incomes necessitates more efficient water-use. Here is an area where both improved governance (or better central administration) complements greater participation (or more effective collective action by citizens). Third, rural people's ability to contribute to the economy must rise through improved health and education. Fourth, agricultural development can make an important contribution to alleviate the employment problem by stimulating the demand for off-farm, labor-intensive goods and services.

Role for the Government

There is consensus that the sustainability of the Egyptian economy, in agriculture, industry, and services, requires a greater role for the private sector. However, it is essential to realize that this does not mean that the government will be or should be given a weaker role. On the contrary, sustainable economic (and agricultural) development requires a stronger public sector. The Government of Egypt needs to stop doing things that the private sector can do better. In this way the government can focus more clearly on performing those functions that it alone can fulfill. Both privatization and decentralization imply greater participation by the citizenry. In the private world of markets and voluntary associations, people make their own choices about how, when, where, and even whether to enter a market or to participate in some form of collective action. Enhanced participation is also necessary if governance is to be more effective: administrators need information about the effects of their activities, e.g., on the impact of an export promotion drive.

The Challenges

Structural adjustment, including strengthened governance and heightened participation, are essential if Egypt is to meet the challenges of the 1990s and coming decades. Egypt faces many challenges, but none are more critical to sustainability than those of food, jobs, human capital formation, and water conservation. In the following pages we will see how structural adjustment, improved governance, and enhanced participation can contribute to meeting these challenges.

Food Security

First, consider the food security picture. The growth of demand for food depends upon the rate of population growth, growth of income per capita, and the income elasticity of demand for food.[2] During the 1980s, Egypt "added a Cairo" to its population: about 12 million additional mouths to feed. By the year 2000 there will be at least 62 million Egyptians, or (nearly) "yet another Cairo." GDP per capita grew annually at just under 3 percent from 1980 to 1989, with most gains occurring in the first half of the decade; the growth rate of GNP per capita was less than 1 percent per year from 1986 to 1990. Should income growth resume, demand for food will surge well above its current rate of increase (about 2.5 percent per year). Supply response was far better during the 1980s than in the 1970s. From 1979 to 1990, Egyptian agricultural output grew at 4.1 percent per year, while food production rose by 6.8 percent per year.[3] Even the growth rate of cereal production (3.0 percent) exceeded population growth. Agricultural production and food production per capita in 1990 were 13 percent and 23 percent, respectively, above that of 1979–1981 (FAO 1990).

This is an impressive performance, but it is no cause for complacency. Consider the fact that, despite the impressive expansion of wheat production—which more than doubled from 1986 to 1990—the cereal self-sufficiency ratio remained essentially unchanged during the decade. Continuing population increase and revived income growth will certainly doom any plans for national self-sufficiency. Egypt's only road to food security lies through diversified, competitive exports of farm and factory products.

Jobs: A Looming Crisis

Egypt needs to create 6 million jobs during the 1990s, simply to keep up with additions to the labor force. If current levels of unemployment are to be reduced, another 1.5 to 2 million jobs must be found. In 1990 the Egyptian labor force was approximately 14–15 million. During the 1990s, roughly 40 percent more jobs must be created just to keep unemployment constant; if unemployment is to be reduced significantly, then the number of jobs must rise by roughly 50 percent. This implies that unless job creation is even more rapid, real wages will stagnate during the coming decade: only if the growth of the demand for labor is more rapid than that of supply can real wages rise sustainably.

This looming employment crisis was foreseeable at least fifteen years ago, which is when today's entrants to the labor force were born. But since the challenge was in the future, little was done. Instead, government employment swelled, spawning artificial jobs at ever lower levels of real remuneration, sacrificing governmental effectiveness at the altar of Nasserist formulas for distributive justice and social stability. Over one-half (55 percent) of all jobs created from 1976 to 1986 were in the public sector (Handoussa 1989). This was not a sustainable employment strategy, as was evident by the mid-1980s. The other source of employment creation, to the crisis of the mid-1980s, was emigration to the Gulf states. Since these jobs were dependent on the oil boom, they, too, could not continue to expand steadily.

Unfortunately, it is relatively unlikely that agriculture can make a significant, direct contribution to employment creation (Richards 1991a). The demand for labor in agriculture is inelastic: recent increases in farm labor supply (as the emigration safety-valve has slammed shut) have mainly driven down real wages, rather than expanded employment. Mechanization has probably proceeded too far for agriculture to play its old role as "shock absorber," disgorging labor in booms and absorbing labor during slumps. By 1990 real agricultural wages had fallen 43 percent from their historical peak of 1985.

Agriculture's main contributions to job creation are likely to be indirect: providing foreign exchange and the stimulus which rising farm income could give to the demand for labor-intensive, rural-industrial goods and services. Some simulations indicate that successful structural adjustment could return Egyptian agriculture to its historical role as a net exporter (Khedr et al. 1989). By raising farmers' incomes, successful structural adjustment could greatly increase the demand for improved housing, furniture, and other labor-intensive goods that could be produced in rural areas—or, at least, in Egypt. Such an "agriculture-development led growth" strategy (Adelman 1984) could be an important complement to the basic export-led strategy, which alone can provide both the foreign exchange and jobs which long-term sustainability requires.

Human Resources

People can contribute to sustainability only if they are educated. One of the most serious inadequacies of public policy during the past generation has been the failure to provide all Egyptians with basic literacy. Only in the late 1980s were all, or nearly all, boys enrolled in primary school; nearly one-fourth of girls are still not enrolled.[4] This legacy will increasingly haunt Egypt. If Egypt must export in order to feed and employ its citizens, it must be able to compete in the international marketplace. Egyptians must produce quality articles and market them successfully. They must be aware of new technologies, and adopt and utilize them effectively. Economic research has shown that, in agriculture as elsewhere in the economy, educated people are more productive. An illiterate workforce is poor material out of which to craft an export-led growth strategy.

Egypt faces strong competition. To produce labor-intensive commodities, it must compete with other low-wage countries and it is instructive to consider their literacy rates. In 1985, when roughly 56 percent of adult Egyptians were illiterate, the corresponding figures for some Asian countries were: Thailand 9 percent; Vietnam 10 percent; Sri Lanka 13 percent; China 31 percent; Malaysia 27 percent; and Indonesia 26 percent (World Bank 1991b). These countries—not Taiwan, Singapore, and Korea, which are already far ahead—are the competition in labor-intensive manufactures in the 1990s. To look closer to the Mediterranean, consider the East European countries (themselves undergoing radical structural adjustment). They have typically universal literacy and a fairly high level of skills. Egypt will have to make strenuous efforts to equip its citizens to meet this intense challenge.

Water-Use: Management by Demand

Egypt's total water supply for the coming decade is essentially fixed at 55–56 billion cubic meters per year. Developments which improve storage and sharing of water among the Nile riparian states will have little positive effect for at least ten years. Indeed, it is easy to foresee negative developments, as, for example, with increased Ethiopian utilization of Blue Nile water resources, or greatly expanded Sudanese irrigation projects. Prudence dictates assuming a fixed supply of water in Egypt for the foreseeable future. It follows that more sophisticated demand management holds the key to optimizing the social value of water. This is basically the responsibility of the agricultural sector: at least 80 percent of all Egyptian water utilization is for crops. The impact of ongoing price reforms for water demand is ambiguous. On the one hand, charging farmers the shadow price of fuel reduces the implicit subsidy to irrigation water. On the other hand, the projected increase in the farm-gate price of cotton should *raise* the demand for water, since cotton is a very water-intensive crop. Good things do not always go together: rationalization of output prices as a whole may raise the demand for irrigation water.

This would not matter much if water were priced. There are, however, serious technical, cultural, legal, and social barriers to explicit water pricing. Some social substitute, which "mimics" pricing, at least to the level of a small group, needs to be sought. The current water allocation system is essentially "supply driven": the central government decides how much water goes where and when. Greater efficiency in water-use implies shifting to a more "demand driven" allocation system, in which farmers' actual needs have a greater influence over allocation. The search for improved water efficiency may dominate agricultural policy discussion in the 1990s and beyond, much as price distortions held the center stage during the 1970s and 1980s.

Structural Adjustment and Politics

The first requirement of political sustainability is the credibility of the structural adjustment process. Unless this works, nothing else will. In general, economists believe that structural adjustment helps the farm sector. Does this mean that farmers are a potential political constituency for structural adjustment? Perhaps, but as we shall see, one must be wary of excessively facile generalizations here.

The reasons for assuming that structural adjustment raises farm incomes are as follows.[5] First, real devaluation raises the price of traded to nontraded goods. Since virtually all farm goods are either traded or have fairly close traded substitutes, real devaluation also improves the farm/nonfarm terms of trade. Within farming, the decline in domestic demand (which is necessary for shor- run stabilization) reduces demand for importables but not for exportables, shifting relative prices toward export crops and away from import competing crops. Real devaluation also may be interpreted as tantamount to a decline in real wages (Harberger 1986). Although there may be some offset (due to the increased profitability of farming), leading to a rightward shift in the demand for labor, this effect (in the Egyptian case) is likely to be more than offset by the autonomous rightward shift in the labor supply function (due to the closure of the "safety-valve" of migration to the Gulf). Real wages will fall.

Structural adjustment in Egypt should raise the income of any farmer who is a net buyer of labor: the impact of real devaluation is reinforced by the decline in labor costs (which were often over half of variable costs in the late 1980s) and by the reform of specific crop pricing policies. Although there may be some counteracting influences from moving input prices to world levels, since purchased inputs are only a fraction of total costs, output price increases will in most cases outweigh such effects. Furthermore, there is evidence that input subsidies did not compensate for output taxation during the 1970s and early 1980s (Dethier 1990). It follows that reform of both should *raise* farm incomes. However, the effects may be dissimilar, depending on the size of farm. Because small farmers depend heavily upon the labor market and sales of dairy products, they are likely to lose from structural adjustment in the short run. The longer-run, general equilibrium effects are, of course, very complex, as output-mixes shift and new technologies are adopted. But in the short run, the conclusion would seem to be that the benefits of structural adjustment are a direct function of farm size.[6]

Participation

The rural constituency for structural adjustment is then potentially considerable: over one-half million (and perhaps as many as one million) farmers stand to gain. Furthermore, Egyptian rural society and politics exhibit many symptoms of "vertical cleavage," of "patron-client" relationships, in which the poorer farmers follow the lead of their wealthier neighbors, upon whom the poor often depend for assistance in times of acute need (Adams 1986). These considerations suggest that there exists a large, relatively undertapped potential constituency for structural adjustment in the countryside. Farmers are, however, relatively unorganized, and their rate of political participation is rather low. Greater participation by farmers in the political process could provide an important offset to the large, well-organized urban constituencies that will suffer short-run losses from structural adjustment.[7]

The contribution by enhanced participation to structural adjustment is not, however, limited to the rural sector. The short-run losers from stabilization and structural adjustment inevitably greatly outnumber the winners: if structural adjustment was necessitated by the country's living beyond its means, it follows that most Egyptians benefited from such profligacy, and will therefore lose from structural adjustment. Structural adjustment also reduces the government's ability to buy off opposition—and recent history (including that of Egypt in the late 1970s) strongly suggests that increased repression will backfire on its user. How, then, to build a constituency for the structural adjustment process and, therefore, for a sustainable economy? One route is to combine political with economic reform, to offer increased participation by those who lose from economic adjustment. To the extent that such groups can be convinced that adjustment is necessary and inevitable, they can be more easily persuaded to negotiate the distribution of the pain. To the extent that they participate in the decisionmaking process, they share the responsibility for the outcome (Bianchi 1990).

Participation is also essential if the private sector is to play an increased role in the economy. The needed private investment for job creation will only occur if Egyptians believe that their property rights are secure. The government's commitment to an expanded role for the private sector must be credible. Creating such credibility requires continued privatization of some industries as well as strengthening the independence of the judiciary. Participation must also increase if the state itself is to act effectively. No bureaucracy can function effectively without feedback from those whom it purportedly serves. Uninformed decisions are usually bad decisions. Greater participation by the citizenry is essential if this information is to be provided. The design of these participation mechanisms will have to be worked out through trial and error; they will probably include a mixture of "voice" (or collective action) and "exit" (market or quasi-market mechanisms) (Hirschman 1970).

Governance

Sustainability requires that the government should stop doing some things, start doing others, and do still other things better. Areas which the government should abandon are well known. Continued government involvement in any productive activity which could become reasonably competitive in private hands is indefensible. Such activities only weaken the state, by spreading it too thinly and diverting resources from other pressing needs. Why should the Government of Egypt lose money on chocolate factories while schools lack even the most rudimentary supplies? Industries which might be oligopolies or even monopolies if privatized (e.g., iron and steel) might well remain in the public sector. However, industry remaining in the public sector needs two things: (1) greater autonomy for managers, specifically including the right to fire workers, and (2) a "hard budget constraint": a credible threat from the central government that it *will not* rescue a failing public company.

Structural adjustment does not weaken the government. A weak government cannot implement structural adjustment. On the contrary, successful structural adjustment requires strong macroeconomic management. Economic reform can succeed only if the government in Egypt becomes more effective in its roles as the tax authority, as the regulator of the banking system, and as the monetary authority. A sustainable agriculture in the coming decades will require a technologically sophisticated, highly responsive, and *strong* central administration over the natural monopoly of the main and branch-canal sections of the irrigation system.

This is but one example of the critical need for government to provide adequate physical and managerial infrastructure (or "hardware" and "software"). Markets cannot work properly if the infrastructure is weak; especially in today's highly competitive international economy, rapid response is required, and this means excellent transportation and communication systems. The Social Fund for Development, with its provision for public works employment for the poor, can combine infrastructural improvements with poverty alleviation. Both contribute to sustainability and require effective governance.

Markets will not work at all if information is absent: none of the neoclassical benefits of markets obtain in situations of poor and highly skewed access to information. Governments have a critical role to play in providing such information to all. Rapid data collection and dissemination and the provision of all types of information is a critical public function that must be greatly strengthened if a sustainable economy is to be created (Klitgaard 1991). In agriculture, it is critical that the government gather and disseminate market information as well as greatly accelerate its research and extension activities. Government must do much more to raise the educational level of the population: this may be thought of as a public investment to raise the information processing capability of the citizenry. Nothing is more vital to Egypt's future.

No matter how well markets function, some citizens will be left behind. The Government of Egypt is committed to providing a social safety net for the very poorest members of society. This is an admirable commitment. Proper functioning of the social fund is an important component of the political sustainability of structural adjustment. The funds have come from abroad, but the management must be local, and it must be effective. Only if the government can quickly identify the poor and swiftly implement programs to provide them with employment or other assistance can the humanitarian and political purposes of the program be realized. The Social Fund for Development could make very important contributions to sustainability, but only if it is properly managed.

Notes

1. Egypt has hardly been alone in such behavior; my own country, the United States, has behaved just as badly and, like Egypt, is now beginning to pay for its previous excesses.

2. That is, $d = n + e(y)$, where d = rate of growth of demand, n = population growth rate, e = income elasticity of demand, and y = rate of growth of per capita income.

3. OLS growth rates calculated from data in FAO production yearbooks. These data confirm the earlier, tentative conclusion in Richards (1991a) that agricultural performance in the 1980s was greatly superior to that of the 1970s.

The differential between food and agricultural production growth may be largely explained by the continued price disincentives to cotton.

4. It should be emphasized here that the current government has made strenuous efforts in this area, efforts that are the more impressive, given the increasingly tight budgetary constraints.

5. The literature is vast. The sketch here follows Norton (1987) and the chapters in Dornbusch and Helmers (1988).

6. A further caveat must be made: some wealthy farmers and agroexporters may enjoy the benefits of past "rent-seeking behavior," and stand to *lose* from structural adjustment. See the examples discussed in Sadowski (1991).

7. An additional important caveat must be made: "cultural politics" (i.e., the debate over the definition and role of Islam) may matter more to people than either economic or political reform.

References

Adams, Richard H., Jr. *Development and Social Change in Rural Egypt.* Syracuse, N.Y.: Syracuse University Press, 1986.

Adelman, Irma. "Beyond Export-Led Growth," *World Development,* 1984, 12: 9.

Bianchi, Robert. "Interest Groups and Politics in Mubarark's Egypt," in *The Political Economy of Contemporary Egypt.* Edited by Ibrahim Oweiss. Washington, D.C.: Georgetown University, 1990.

Dethier, Jean-Jacques. *Trade, Exchange Rates and Agricultural Pricing Policies in Egypt.* Washington, D.C.: World Bank, 1990.

Dornbusch, Rudiger, and C.H. Helmers. *The Open Economy: Tools for Policymakers in Developing Countries.* New York: Oxford University Press, 1988.

FAO. *Production Yearbook.* Rome: FAO, 1990.

Handoussa, Heba. "The Burden of Public Sector Employment and Remuneration: A Case Study of Egypt." Geneva: ILO, 1989.

Handoussa, Heba, and Gillian Potter. Editors. *Employment and Structural Adjustment: Egypt in the 1990s.* Cairo: American University in Cairo Press, 1991.

Harberger, Arnold C. "Applications of Real Exchange Rate Analysis." Mimeo. 1986.

Hirschman, Albert O. *Exit, Voice, and Loyalty: Responses to Decline in Firms, Organizations, and States.* Cambridge, Mass.: Harvard University Press, 1970.

Khedr, Hassan, Leroy Quance, and Bruce McCarl. "Evaluation of the Egyptian Agricultural Sector: Implications for Further Decontrol." Dokki: Ministry of Agriculture Working Paper No. APAC-89-8, 1989.

Klitgaard, Robert. *Adjusting to Reality: Beyond "State Versus Market" in Economic Development.* San Francisco: International Center for Economic Growth, 1991.

Norton, Roger D. "Agricultural Issues in Structural Adjustment Programs." FAO Economic and Social Development Paper. Rome: FAO, 1987.

Richards, Alan. "Agricultural Employment, Wages and Government Policy During and After the Oil Boom," in *Structural Adjustment: Egypt in the 1990s.* Edited by H. Handoussa and Gillian Potter, 1991a.

———. "The Political Economy of Dilatory Reform: Egypt in the 1980s," *World Development,* 1991b, 19: 12.

Sadowski, Yahya M. *Political Vegetables? Businessman and Bureaucrat in the Development of Egyptian Agriculture.* Washington, D.C.: Brookings Institution, 1991.

World Bank. *Poverty Alleviation and Adjustment in Egypt.* Washington, D.C.: World Bank, 1990.

———. *Structural Adjustment Loan Document: Egypt.* Washington, D.C.: World Bank, 1991a.

———. *World Development Report 1991.* New York: Oxford University Press, 1991b.

Conference Organizing Committee

Chairman:

Dr. Youssuf Wally,
Deputy Prime Minister and
Minister of Agriculture and Land Reclamation,
Government of Egypt, Cairo, Egypt

Secretary:

Dr. Mohamed A. Faris,
Professor, Faculty of Agricultural and Environmental Sciences,
McGill University,
and Director, CEMARP, Montreal, Canada

Members:

Dr. Roger Buckland, Vice Principal of McGill University and
Dean, Faculty of Agricultural and Environmental Sciences,
McGill University, Montreal, Canada

Dr. Ahmed Momtaz,
Advisor for Research Affairs,
Ministry of Agriculture and Land Reclamation,
Government of Egypt, Cairo, Egypt

Dr. Mohamed A. Sabbah,
Dean, Faculty of Agriculture,
Alexandria University, Alexandria, Egypt

Mr. Aly Shady,
Chief, Irrigation Sector,
Canadian International Development Agency,
Hull, Quebec, Canada

Conference Participants

Dr. Adel Mahmoud Aboul-Naga, Undersecretary for Animal Production, Ministry of Agriculture and Land Reclamation, Cairo, Egypt

Dr. Kheiry Aboul-Seoud, Chairman, Department of Agricultural Extension and Rural Sociology, Cairo University, Cairo, Egypt

Dr. Mahmoud Abu-Zeid, Chairman, Water Research Center, Ministry of Public Works and Water Resources, Cairo, Egypt

Dr. Hoda Badran, Secretary General, National Council for Childhood and Motherhood, Cairo, Egypt

Dr. Osman Adly Badran, Former Minister of Agriculture, Government of Egypt, and Emeritus Professor, Alexandria University, Alexandria, Egypt

Dr. Mohamed Sayed Balal, Director, Rice Research and Development Program, Agricultural Research Center, Ministry of Agriculture and Land Reclamation, Cairo, Egypt

Dr. C. Fred Bentley, Emeritus Professor, University of Alberta, Edmonton, Canada

Dr. Adli Bishay, Director General, Desert Development Center, the American University, Cairo, Egypt

Dr. Fahmy K. Bishay, Chief, Policy and Planning Service, Policy Analysis Division, FAO, Rome, Italy

Dr. Asit K. Biswas, International Development Centre, Oxford University, Oxford, United Kingdom

Dr. Roger Buckland, Dean, Faculty of Agricultural and Environmental Sciences, Macdonald College, McGill University, Montreal, Canada

Dr. Michael M. Cernea, Senior Advisor, Environment Department, World Bank, Washington, D.C., United States

Dr. Bruce E. Coulman, Chairman, Department of Plant Science, Macdonald College, McGill University, Montreal, Canada

Dr. Hamdy M. Eisa, Principal Agriculturist, Agriculture and Rural Development Department, World Bank, Washington, D.C., United States

Dr. Mohamed Abou-Mandour El-Deeb, Professor of Agricultural Economics, Faculty of Agriculture, Cairo University, Cairo, Egypt

Dr. Osman Ahmed El-Kholie, Economic Advisor to Minister of Agriculture and Land Reclamation, Government of Egypt, Cairo, Egypt

Dr. Hosny El-Lakani, FAO Regional Office for the Near East, Cairo, Egypt

Dr. Nabil Mohammed El-Moelhi, Director, Soil and Water Research Institute, Ministry of Agriculture and Land Reclamation, Cairo, Egypt

Dr. Abdel Tawab Amin El-Mohandes, Emeritus Professor, El-Menia University, El-Menia, Egypt

Dr. Salah El-Serafy, Senior Advisor, Environment Department, World Bank, Washington, D.C., United States

Dr. Mohamed Fawzy El-Shaarawy, Vice-President, Ain Shams University, Cairo, Egypt

Dr. Khaled El-Shazly, Emeritus Professor, Department of Animal Science, Alexandria University, Alexandria, Egypt

His Excellency *Dr. Farouk El-Telawi,* Governor, The New Valley Governorate, El-Kharga, Egypt

Dr. Salah M. El-Zoghby, Professor and Chairman, Department of Rural Sociology, Faculty of Agriculture, Alexandria University, Alexandria, Egypt

Dr. Mohamed A. Faris, Professor, Faculty of Agricultural and Environmental Sciences, McGill University, and Director, CEMARP, Montreal, Canada

Dr. Mohamed Maamoun Abdel Fattah, Advisor to Ministry of Finance, Government of Egypt, Cairo, Egypt

Dr. Mazhar Fawzy, Professor of Agronomy, Cairo University, Cairo, Egypt

Dr. Sayed Galal, Jr., Emeritus Professor of Agronomy, Cairo University, Cairo, Egypt

His Excellency *Dr. Ahmed A. Goueli,* Governor, Ismailia Governorate, Ismailia, Egypt

Mr. Salah Hafez, First Undersecretary, Environmental Affairs Organization, Ministry for Environmental Affairs, Cairo, Egypt

Dr. Atef Hamdy, Director of Research, International Centre for Advanced Mediterranean Agronomic Studies, Bari, Italy

Dr. Stuart B. Hill, Director, Ecological Agriculture Projects, Macdonald College, McGill University, Montreal, Canada

Dr. Nicholas S. Hopkins, Department of Sociology and Anthropology, The American University, Cairo, Egypt

Dr. Fouad H. Khadr, Professor and Chairman, Department of Agronomy, Faculty of Agriculture, Alexandria University, Alexandria, Egypt

Dr. Mahmood Hasan Khan, Professor of Economics, Simon Fraser University, Vancouver, Canada

Dr. Fawzy Kishk, Regional Director, International Development Research Centre (IDRC), Cairo, Egypt

Dr. Angus F. MacKenzie, Faculty of Agricultural and Environmental Sciences, Macdonald College, McGill University, Montreal, Canada

Dr. Ahmed Momtaz, Advisor for Research Affairs, Ministry of Agriculture and Land Reclamation, Cairo, Egypt

Dr. Ahmed Mostagir, Dean, Faculty of Agriculture, Cairo University, Cairo, Egypt

Dr. Ahmed Tahir Moustafa, Deputy Director, Soil and Water Research Institute, Ministry of Agriculture and Land Reclamation, Cairo, Egypt

Dr. Ngozi Okonjo-Iweala, Chief of Agricultural Operations, World Bank, Washington, D.C., United States

Dr. Bakir A. Oteifa, Advisor to Minister of Agriculture and Land Reclamation, Government of Egypt, Cairo, Egypt

Mr. Joseph R. Potvin, Consultant, Economy, Community, Environment, Ottawa, Canada

Dr. Mohamed Abd El-Hady Rady, Chairman, Egyptian Public Authority for Drainage Projects, Ministry of Public Works and Water Resources, Cairo, Egypt

Dr. Alan Richards, Professor of Economics, University of California, Santa Cruz, California, United States

Dr. Mohamed Ahmed Sabbah, Dean, Faculty of Agriculture, Alexandria University, Alexandria, Egypt

Mr. Aly M. Shady, Chief, Irrigation Sector, Natural Resources Division, Canadian International Development Agency, Hull, Quebec, Canada

Dr. Kamilia Mohammed Shoukry, Consultant, National Council for Childhood and Motherhood, Cairo, Egypt

Dr. Ismail Sirageldin, Professor of Population Dynamics and Economics, Johns Hopkins University, Baltimore, Maryland, United States

Dr. Greg Spendjian, Acting Director, Agriculture, Food, and Nutrition Sciences Division, International Development Research Centre (IDR), Ottawa, Canada

Dr. Howard A. Steppler, Emeritus Professor of Agronomy, Macdonald College, McGill University, Montreal, Canada

Dr. Samir Toubar, Vice-Chancellor and Professor of Economics, Zagazig University, Zagazig, Egypt

Dr. E.T. York, Jr., Chancellor Emeritus, Florida State University System, Florida, United States

Index

About the Book

Egypt's agricultural development has been constrained by, among other factors, the need to conserve scarce natural resources, the pressures of rapid urbanization, the onslaught of the desert, and, not least important, technological limitations and restric- tive economic structures. This book addresses the issues crucial to achieving and maintaining sustainable agriculture in Egypt.